CAMBRIDGE EARTH SCIENCE SERIES

Editors: W. B. HARLAND (*General Editor*), S. O. AGRELL
A. H. COOK, N. F. HUGHES

Palaeobiology of angiosperm origins

IN THIS SERIES

MINERALOGICAL APPLICATIONS OF CRYSTAL FIELD THEORY
by R. G. Burns

NON-MARINE ORGANIC GEOCHEMISTRY
by Frederick M. Swain

PALAEOMAGNETISM AND PLATE TECTONICS
by M. W. McElhinny

JURASSIC ENVIRONMENTS
by A. Hallam

ATLAS OF QUARTZ SAND SURFACE TEXTURES
by David H. Krinsley and John C. Doornkamp

Palaeobiology of angiosperm origins

Problems of Mesozoic seed-plant evolution

NORMAN F. HUGHES

University Lecturer in Geology
Fellow of Queens' College, Cambridge

CAMBRIDGE UNIVERSITY PRESS

CAMBRIDGE

LONDON · NEW YORK · MELBOURNE

CAMBRIDGE UNIVERSITY PRESS
Cambridge, New York, Melbourne, Madrid, Cape Town, Singapore,
São Paulo, Delhi, Dubai, Tokyo

Cambridge University Press
The Edinburgh Building, Cambridge CB2 8RU, UK

Published in the United States of America by Cambridge University Press, New York

www.cambridge.org
Information on this title: www.cambridge.org/9780521287265

First published 1976
Re-issued in this digitally printed version 2010

A catalogue record for this publication is available from the British Library

Library of Congress Cataloguing in Publication data

Hughes, Norman Francis.

Paleobiology of angiosperm origins.

(Cambridge earth science series)

Includes bibliographical references and index.

1. Angiosperms, Fossil. 2. Palaeobotany–Mesozoic. I. Title.

QE980.H8 561'.2 75-3855

ISBN 978-0-521-20809-3 Hardback
ISBN 978-0-521-28726-5 Paperback

Contents

Contents

Preface

Study of this long-standing problem is just beginning to take a new and more promising turn. The great development in the last two decades of knowledge of fossil pollen has made possible a truly interdisciplinary approach from both geology and botany, whereas previously the work put into this subject had been mostly botanical. This contribution aims both to analyse method and purposes closely, and to encourage faith in the adequacy of the fossil record for making possible a continuing and effective solution.

Although none of them bear any responsibility for what is here offered, I believe that much appreciated encouragement both past and present has come from the works of or from personal contact with W. T. Gordon, H. Hamshaw Thomas, W. N. Edwards, Leonard Hawkes, Maurice Black, Tom M. Harris, Sir Harry Godwin, Marjorie E. J. Chandler, Chester Arnold, Armen Takhtajan, Henry N. Andrews, Harlan P. Banks, J. M. Schopf, Richard A. Scott, Percival Allen, W. B. Harland, Kenneth Sporne, John Smart, James Doyle, W. G. Chaloner, and V. A. Krassilov. Others whom it would be most unfair to mention individually have stimulated, mostly without their knowledge, by opposition which should never be under-estimated.

I acknowledge with gratitude much technical assistance from Mrs M. G. Parmée, Miss C. A. Croxton, Miss Christobel King, Mr John Lewis and others of the Department of Geology, Cambridge. I am grateful to the Trustees of the British Museum (Natural History) for permission to use material from which figures 8.7, 8.8, 8.9, 8.10, 8.11 and 8.12 have been adapted.

N.F.H.

1 Introduction and proposition

The purpose of writing this book is to present in a new light the old problem of the evolutionary origin of the angiosperms, and particularly to suggest that in spite of the difficulties frequently reported a solution is entirely possible.

THE PROBLEM

The evolutionary origin of the now dominant land-plant group, the angiosperms, has puzzled scientists since the middle of the nineteenth century. In late Cretaceous rocks angiosperm fossils were dominant, but in the early Cretaceous seed-plant fossils were almost entirely of gymnospermous types; the transition appeared to be sudden. On account both of the regular nature of the structure of the angiosperm flower and of the remarkable double fertilisation detail, it had for long been assumed that angiosperms were monophyletic and therefore that the immediate ancestor should have been traceable among the relatively few known fossil and Recent gymnosperm groups. None of these gymnosperm groups however appeared to carry a set of characters which was at all near to the requirement. Palaeobotany did not appear to provide clarification, and botanists turned instead to hypothetical constructions of the 'primitive flower'. These could not be tested and consequently in due course interest lapsed.

As evolutionary studies in fossil animals became relatively well documented, three interdependent 'escape' hypotheses were developed to explain, in the plant kingdom, the continuing failure to deal with the angiosperm 'mystery'. These were: (*a*) general incompleteness of the plant fossil record in addition to the admitted lack of fossils of the flowers themselves; (*b*) slowness of evolution in land plants calling for a long developmental history to produce such a complication as angiospermy; and (*c*) cryptic evolution of early angiosperms in upland areas from which few if any fossils could have been preserved. Although all these hypotheses were depressingly negative, palaeobotanical search continued and still does. With few exceptions of detail however the failure to find a satisfactory explanation has persisted and many botanists have concluded that the problem is not capable of solution by use of fossil evidence; geologists have displayed little interest in the matter, apparently owing to a general lack of coherence of philosophical treatment of palaeobiological information.

Behind the original botanical enquiry was the need to improve the basis of classification of the living angiosperms which still form the largest group of organisms without an agreed or satisfactory hierarchical arrangement of taxa. New information from such diverse sources as numerical taxonomy, chromosome numbers, palynology and chemotaxonomy appeared to complicate rather than to clarify the position in a most extensive literature. Analysis of characters as either primitive or advanced has helped, but although classifications in use appear to have improved, large numbers of doubts and anomalies remain in them. A high proportion of current work goes directly in the widest sense into comparative morphology of Recent taxa without any historical perspective.

In general at this time botanists have turned away understandably from this mystery which does not appear to 'crack' while such great advances are made in other botanical fields. This has probably influenced the apparent contemporary decision in botany to devote little time and resources to the study of either palaeobotany or taxonomy. Regrettable as this is on general philosophical grounds, some justification is clearly needed for recommending any change.

PROPOSITION

Following from general consideration of the literature and some detailed study of fossils over the last twenty-five years, and from various attempts to approach the problem from different directions, my proposition is that there is nothing unusual about this evolutionary problem, and that the lack of progress has mainly been caused by the nature of certain biological and botanical prejudices and by relative palaeontological (and thus geological) neglect. It is consequently necessary to examine first the situation in the science and then all the relevant fossils before predicting in chapter 12 how progress is likely to be made; at this stage also it will consist more of indicating direction and possibility rather than of claiming any detailed solution.

By concentrating on the fossil record itself and on how it may be used, and by dismissing unprofitable lines of enquiry such as imaginary 'upland' fossil floras, a true systematic search will become possible. This should in time logically elucidate both the origin and the early evolution of the angiosperms without any recourse to special theory. It should also provide eventually a basis for a true phylogeny leading to the numerous living representatives of the angiosperms, but this will be one in which *the floral parts of the plant will play only a subordinate role.* Whether this unusual basis will be allowed to affect classification of the extant organisms remains to be seen.

EVOLUTION OF THOUGHT

It is desirable if possible to forestall a criticism that this proposition is a brash rejection, based only on relative ignorance, of much former work. An answer is my belief that all honest scientific work has its time, before which it is not conceptually possible and after which it will eventually be superseded.

PART 1

Technical situation

2 Biology and Earth evolution

Before entering more specialised and detailed discussions it is almost certainly necessary to clarify some general assumptions about convention and about certain biological problems, which will be made here throughout but with no further comment.

2.1 EARTH EVOLUTION

Biologists seldom appear to allow for the whole earth (and beyond) being concerned in continuous physical and chemical evolution, of which the much discussed 'biological evolution' is only a sector. Perhaps geologists also have been slow to point out what advances they have made in recent years.

In the last decade traditional studies on animal fossils of the 600 million years of Phanerozoic time have been supplemented by work on aquatic plant fossils ranging back over 2000 million years; this work has brought a clearer realisation that the characters of the atmosphere, hydrosphere and sedimentary processes of the earth's crust have been throughout (and of course still are) controlled by the presence and activities of living organisms as a whole. The hypothesis of crustal plate tectonics and the study of lunar rocks have correspondingly reminded geologists of the evolutionary nature of all the differentiation and concentration processes of the high-temperature earth beneath the crust.

Palaeobiologic evolution can thus be seen as a small but to man an important part of an all-embracing process of which it is most difficult to study any part successfully in the isolation of any one traditional discipline.

2.2 'UNIFORMITARIANISM'

It is difficult to study and reconstruct past processes, especially in the knowledge that they represent a succession of events each to some extent dependent on previous events. To enter such a field it is necessary to analyse relevant current processes in order to identify the types of parameter or factor which could have been involved in the past. As long as the slogan 'the present is the key to the past' refers only to such an analysis, the procedure is sound; however, as soon as it is extended, and called 'uniformitarianism', it is usually taken to imply such a close parallel of circumstances between present and past that the evolutionary elements become overlooked. In the context of the current problem this unimaginative and certainly inaccurate procedure has been applied elsewhere to various types of Cretaceous sedimentation and to the mode of life of Cretaceous gymnosperms; some stress will therefore be laid on avoiding these pitfalls.

2.3 NATURAL SELECTION

The only control on evolution subsequent to the origin of life has been natural selection, for which the term 'geological event selection' is preferred. This preferred term is intended to include as well as physical environment changes the more static chemical accidents which are local in space but related to past environment changes. Even what appears to have been pure biological 'competition' was probably based on geological environment changes.

The use of the words 'adaptation' and 'adaptive' is consequently avoided throughout as unnecessary because every character of an organism falls in this category. A 'non-adaptive' character is a character not properly understood. The word 'pre-adaptive' belongs to journalism after the event.

2.4 PALAEOBIOLOGY/GENETICS INTERFACE?

If a Mesozoic ammonite zone is estimated to be of approximately two to five hundred thousand years duration, and if some palaeobiological correlation under favourable circumstances may be a little finer, there must remain a large gap either in generations or in years between this and any genetic experience. This means that from a palaeontologic point of view all the fossil assemblages (or in the most favourable cases, populations) may exhibit a gross variation among individuals, but it cannot yet be meaningfully discussed in terms of genetics.

It is not yet possible for example to assess the importance of polyploidy in the past; although it probably has considerable influence now in taxonomy of ferns and of herbaceous angiosperms, particularly from high latitudes, it has not been recorded at all in living gymnosperms; it may therefore have originated in the Tertiary with the spread of herbaceous angiosperms and of the higher leptosporangiate ferns.

2.5 SYSTEMATICS AND PALAEOBIOLOGY

Hitherto in almost all palaeobiological work, the method has been to copy and perhaps extend the organisation for systematic treatment of living organisms. This has been followed in all taxonomy of both animal and plant fossils, often without much special thought concerning nomenclature and even sometimes to the extent of editorial enforcement. For various reasons this procedure has proved inefficient and also an actual hindrance to data-handling and interpretation. Some alternative method for handling fossils is considered essential.

2.6 FUNDAMENTAL TAXA IN SPACE AND TIME

It has been suggested by Joysey (1956) and by others that there is a useful equivalence between a living species with its agreed spatial limits on one time-plane and a palaeontologic taxon with its limits in time; the parallel has even been assumed to give support to the custom of using the same systematic methods for the two sets of data. There is however no parallel between (*a*) the actual genetic discontinuity in space on one time-plane between living species which can be distinguished, and (*b*) the theoretical genetic continuity of a succession of generations in time from which any taxa erected can only be arbitrarily delimited and defined. This clear difference is the basis of the suggestion in chapter 4 below that palaeontological data-handling requires an entirely separate code of practice.

2.7 'BACKWARDS SYSTEMATICS'

It is logically essential to avoid backwards systematics which is the persistent use of living taxa of all grades to accommodate Palaeogene and Mesozoic fossils. Virtually all fossil angiosperms have been treated in this way (see Chesters *et al.* 1967), although the procedure completely precludes meaningful discussion of evolution within the membership of the families or other groups concerned (fig. 2.1). How can the details of descent of a living plant taxon from some Cretaceous fossils be determined, when the latter have already been identified with the former?

To study Jurassic plants it is wise first to look at Carboniferous to Triassic floras to see whence the plants may have evolved. Unfortunately a large and influential sector of fossil plant study in the Quaternary has been deliberately organised backwards for stratigraphic purposes, and the existence of this sector has encouraged extension of this illogical procedure; Quaternary plant fossils are identified into living taxa only on the excuse that the time-span of that period is too short for evolution to be detected. Practical as such procedure may be in certain circumstances, its use should clearly be restricted to the smallest possible sector of geologic time.

As will be discussed in chapter 4 another way of looking at this practice is to regard it as an entirely unwarranted extrapolation. As an example of this, several well-known palaeobotanists have preferred to classify certain Jurassic fossil compressed leaves as *Ginkgo* Linn. or even as *G. biloba* L. Other parts of the fossil plants than leaves are not certainly known from the same beds nor even from the same area; the leaves are of the same shape as those of the single living species and the cuticle is similar but not identical. Such an attribution is unlikely to be made in the future if the true purpose of the fossil classification is recognised.

2.8 ADDITION OF NEW INFORMATION

In any system of organising and using descriptive data there will be a problem of incorporating information from new exploration or from new techniques into the working diagnosis of existing taxa. So far in palaeontology this kind of 'emendation' activity has been conducted solely on the basis of the individual specialist's expert memory or filing system; it has become increasingly difficult for anyone else to handle the data effectively. Now that greatly extended memory storage is available or likely soon to be so, the problem becomes the 'mathematical' one of presenting the information usefully and in such a way that new data can be added continuously and used with increasing effects by the non-specialist.

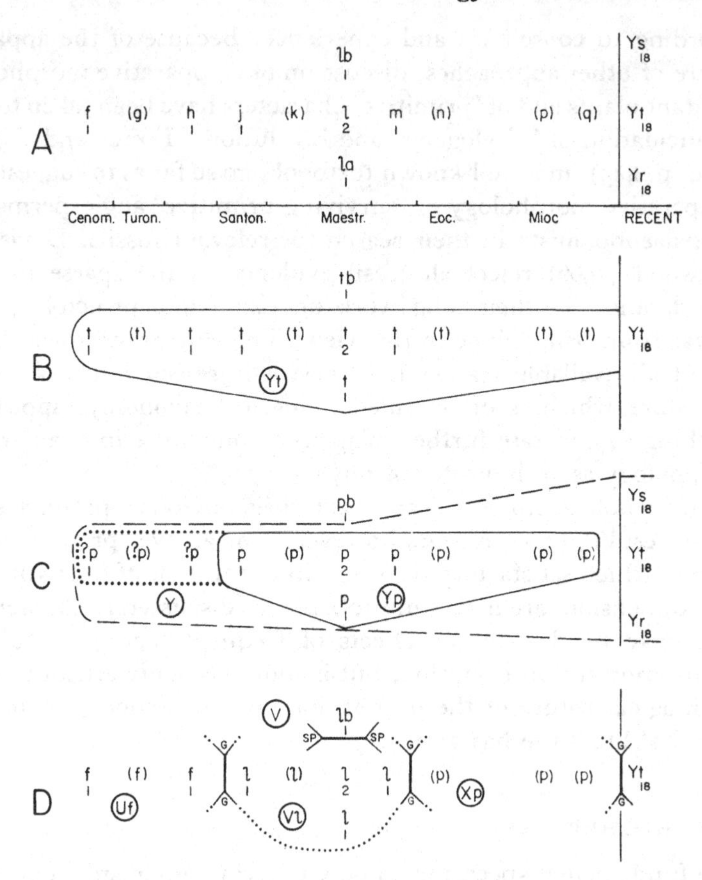

Figure 2.1. Diagrams to illustrate the disadvantages of 'backwards-systematics' for stratigraphic/evolutionary studies. A set of data stratigraphically arranged; *f*, *h*, *l*, etc. represent fossils with their number of available characters indicated below; (*g*), (*k*) etc. represent unseen fossils of similar kind known from other areas, with no definite number of characters shown. B the effect of identifying most of the fossils with the living species *Yt*. C the effect of using a fossil species, but an extant genus *Y*. D erection of taxa for fossils with both time and morphologic limits defined where possible, with the potentiality for continuing the process. Using D, unlike B and C, it is possible to store, retrieve and discuss geological events (data) independently and for all geologists with access to the information to study the course of evolution.

2.9 COMPARATIVE MORPHOLOGY ON ONE TIME-PLANE

Systematic botanists in this field are faced with the problem of handling the very large angiosperm group by submitting it to their classifications that are labelled artificial, natural or phylogenetic according to conscience and experience. Because of the apparent failure of other approaches, discussion of comparative morphology of extant plants and of 'primitive' characters have been taken to lead to elucidation of 'phylogeny' and 'evolution'. Foster and Gifford (1959, p. 445), in a well-known textbook, go so far as to suggest that comparative morphology of 'surviving primitive' angiosperms will aid palaeobotanists in their search for relevant fossils. Davis and Heywood (1966) reject all fossil evidence as too sparse for any consideration in their long work on taxonomic principles; they are far from being alone in this view. The slow trend towards the use of all available (rather than narrowly selected) characters, a procedure which is often called numerical taxonomy, appears if anything to generate further misplaced confidence in comparative morphology as an indicator of phylogeny.

This whole attitude and approach does obviously produce some useful results; there appears however to be an ever-present cut-off beyond which satisfactory results, without the use of fossils and the time dimension, are highly improbable (as discussed in chapter 14). This may be due to the effects of frequent 'convergence' and 'homeomorphy' in evolution, but is more certainly attributable to the illogical nature of the proposition on which such an extensive botanical literature has grown.

2.10 HIGHER TAXA

The fundamental species taxonomy of living organisms remains at present relatively stable in concept, whereas because of the time dimension there is no really valid species concept or taxon stability in palaeobiology.

The great wisdom behind the Linnean binominal nomenclature is the obligation on extant taxon originators to classify once into a genus and thus to state distinction criteria against adjacent taxa. This kind of obligation is also valuable in a palaeontological scheme.

Higher taxa than the genus are merely language and filing devices and they justify themselves on the one (Recent) time-plane by their utility. When they are applied to fossil information, it is important to include the idea that the framework of higher taxa must change with time, and must be flexible enough to be meaningful and useful on each time-plane that is involved. This aspect of the usual hierarchical arrangements is discussed in chapter 16.

2.11 DISTINCTIVE PURPOSE IN PALAEOBIOLOGY

The sole economically justifiable purpose of the relatively labour-expensive and esoteric pursuit of palaeobiology is the elucidation of earth evolution, otherwise 'Stratigraphy' in its broadest definition. The palaeobiologic sector of earth evolution remains by far the most effective basis for stratigraphic time-correlation in the sense that events may be selected and then ordered (correlated) in time-sequence.

Because a palaeobiologic species must contain the individuals of an arbitrarily chosen number of generations, and because each of its time boundaries must always separate a parent from its progeny, it has no natural unity or distinction from its neighbour in time. It is therefore merely a palaeontologist's selection rather than any natural entity, and as such should be made to suit a defined purpose. This purpose should be stratigraphic utility, and the procedures of taxonomy and nomenclature surrounding it should reflect and assist that decision.

Although palaeobiology relies properly on biology of living organisms for initial training purposes, palaeontologists have been slow and reluctant to break away and to set up their own procedures and codes of practice for the handling of their very different material and information. This reluctance has almost certainly been the cause to date of the relative isolation and ineffectiveness of much palaeontological work.

If palaeobiological information can be reorganised on a more effective and purposeful basis, the results will nowhere be more strongly felt and far reaching than in the field of classification of living angiosperms.

3 Palaeobotanical factors

The study of palaeobotany is unfortunately a less familiar discipline than palaeozoology and it has often been totally neglected because of its supposed difficulty. There are some peculiarities in the study of land plants although they are likely to occur in some measure in the study of other fossil material.

3.1 HISTORICAL DEVELOPMENT

In the nineteenth century palaeobotany appears to have been widely studied, but in the period 1920–30 some unfortunate decisions appear to have been taken with the effect of restricting both teaching and available finance in palaeontology to marine invertebrate palaeozoology because it alone then showed more stratigraphic promise. In the Carboniferous Coal Measures in Britain, megafossil palaeobotany did fail in the early part of this century to produce adequate stratigraphic results when they were needed at a critical time, and the consequent reaction can be understood. The misfortune is that palaeobotany has now been left in a neglected position at a time when a genuinely integrated effort in every available palaeontological field is needed. The rise of palaeopalynology has begun to correct this imbalance but with these microfossils there are serious data-handling dangers that are discussed below.

3.2 PRESERVATION OF LAND-PLANT FOSSILS

What may be termed the soft parts are more easily and commonly preserved in plants than in animals, because virtually all land-plant tissue possesses some rigidity however slight; the form of the individual cells can thus be preserved although only rarely is there any trace of the cell contents; plant cell walls are normally of cellulose but in certain organs they may be too delicate to be preserved. Cell walls of wood tissues have considerable thickening

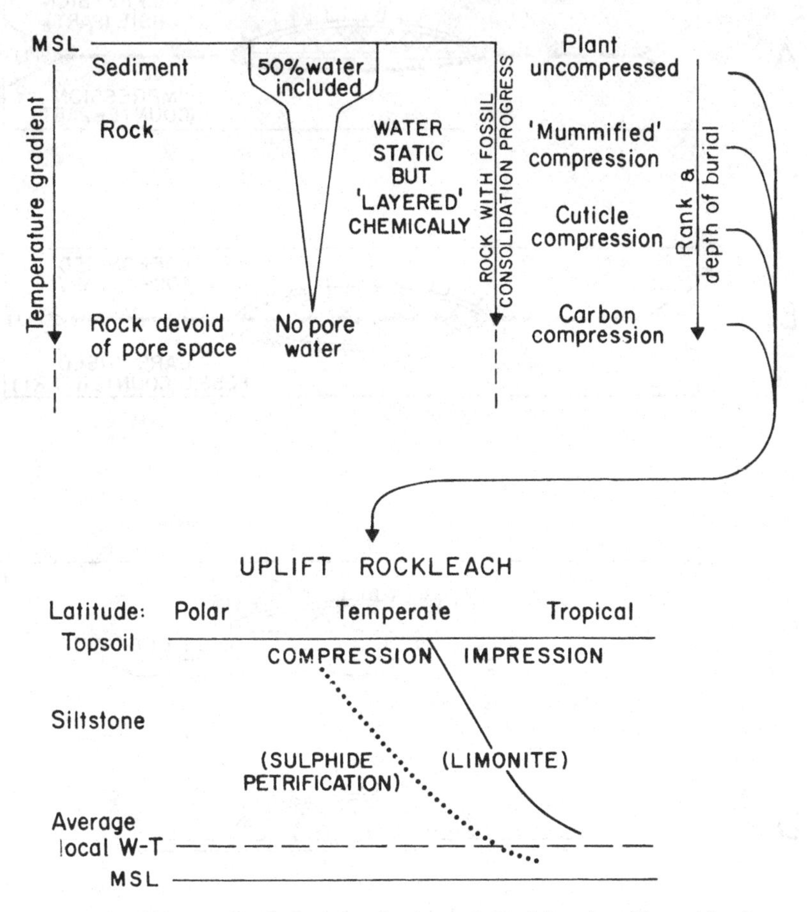

Figure 3.1. Diagenesis of plant fossils through burial and uplift rockleach. MSL = mean sea-level; W-T = water-table.

of lignin which is much more rigid, and is more durable, although not in terms of geological time. However, by far the most inert of cell-wall materials in plants are the cutin-covered surfaces of stems, leaves and seeds, and the sporopollenin in the walls of spores and pollen grains; the function of such materials in life was primarily the occlusion of water and of the solubles in it, and the resulting diagenetic products of them seem to last indefinitely when they have not been subjected to oxidation; they are consequently the basis of the majority of fossil information.

Most plant fossils are compressions, which in essence usually

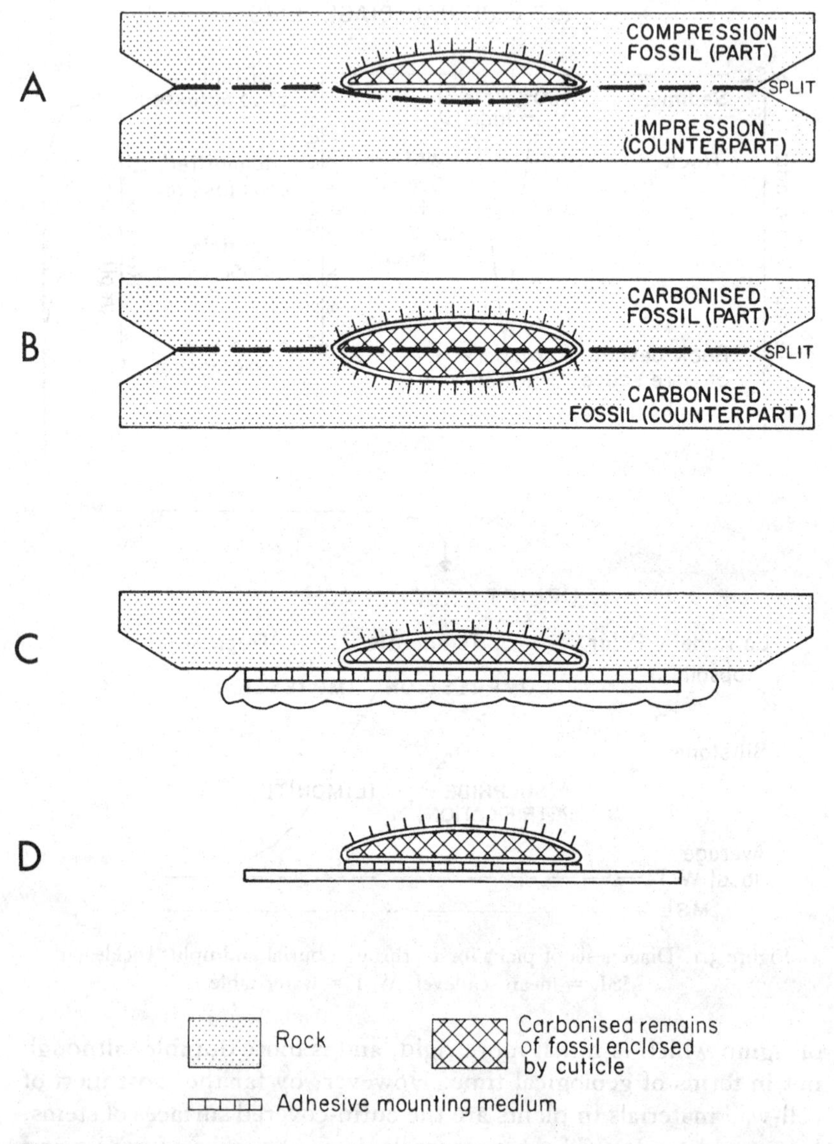

Figure 3.2. Compression: preservation details. A typical rock split revealing cuticle compression; upper cuticle with projections (hairs etc.). B typical rock split providing two equal fossils, part and counterpart; upper and lower cuticle with projections. C fossil of either type mounted with adhesive on glass plate. D completed 'transfer' preparation, with cuticle surface exposed in detail; rock has been dissolved away.

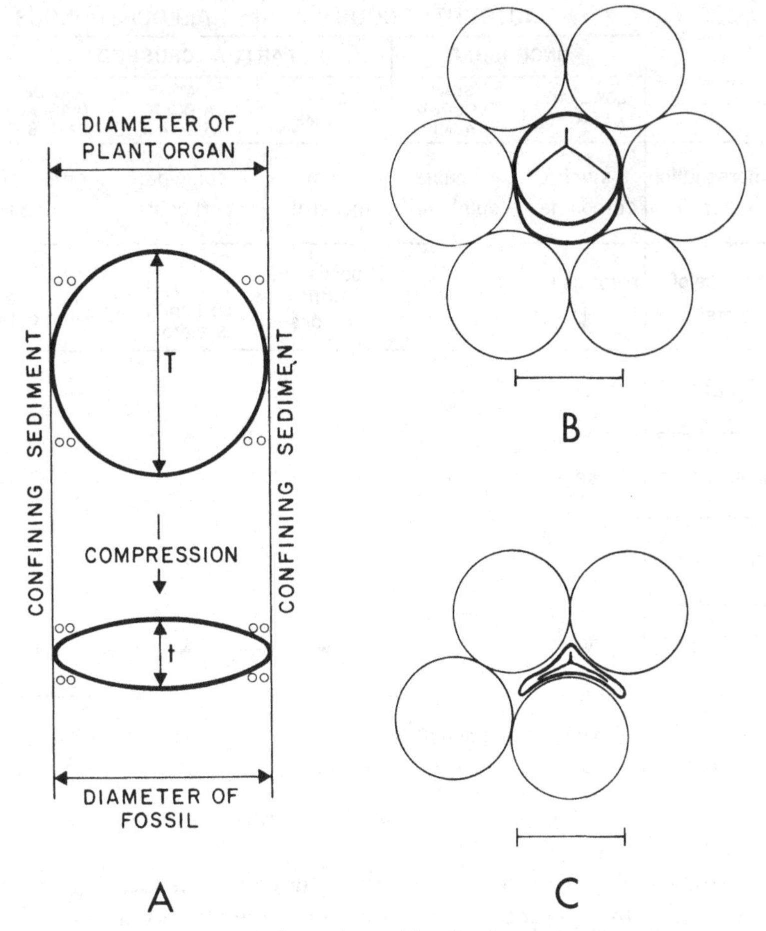

Figure 3.3. Compression, and dimensions of fossils. A megafossil gravity compression; reduction of vertical thickness (*T*, *t*) with unchanged diameter. B microfossil (miospore) in unconsolidated wet sediment. C miospore compressed in consolidated sedimentary rock with close-packing of grains, diameter of miospore unchanged.

consist of the consolidated opaque carbonised remains of cellulose and lignified cell walls enclosed within a translucent sheath of cutin or sporopollenin. Although embedded in sedimentary rocks, such fossils are clearly subject to diagenetic consolidation processes, and later to 'rockleach' effects if the surrounding rocks are uplifted and are relatively permeable (fig. 3.1). They may further be subject to extraction damage if the cuticle bears sufficient small processes for its adhesion to the adjacent rock to be strong. This may cause rock

	AUTOCHTHONOUS			ALLOCHTHONOUS	
	UNCRUSHED		PARTLY CRUSHED		
	LOWLAND PEAT	SEA-FLOODED SWAMP	COAL SWAMP	AT SEA-BOTTOM INTERFACE	IN DIAGENESIS AFTER BURIAL
Impregnation material	silica; carbonate	carbonate; sulphide	coal; maceral	sulphide; carbonate	carbonate; silica
Source of material	pyroclastics ↓		impersistent plant organs	sulphur bacterial, carbonate skeletal	↑ limestone or pyroclastics
LEAF	L	L			
SPORE/POLLEN	SP	SP	SP		
FRUIT/SEED	F	F		F	
WOOD	W	W	W	W	W
Example flora	Rhynie chert	Coal-ball floras	Jet, coal	London Clay	Arizona Triassic

Figure 3.4. Origin of plant petrifactions.

splitting to fall within the fossil rather than along its face (fig. 3.2*B*); splitting within the fossil does however leave the potentiality for a chemical transfer preparation of all the delicate details of a cuticular surface (fig. 3.2*D*). Rendering an opaque fossil translucent for study of the detail of exine or cuticles requires strong oxidants to make soluble for removal the semi-carbonised and carbonised remains of the cell contents and any persisting fragments of cellulose and lignin cell walls; it is sometimes difficult to arrange this oxidation without incidentally damaging the cuticle, and damage of this kind may well produce SEM (scanning electron microscope) artefacts. A general question which arises with compressions concerns the dimensions of the fossil in the plane of compression; these dimensions are measured and referred to as if they were equal to original horizontal diameters (fig. 3.3); this important assumption normally appears to be justified (but see Harris 1974).

A small number of plant fossils are impregnation petrifactions in which mineral (calcium carbonate, sulphides, or silica) or coal maceral has filled the cell lumina; all or part of the original cell-wall material remains between the mineral-filled lumina, and thus there is sufficient contrast for the fossils to be studied in detail by thin rock-section or by peel. This preservation occurs just after burial or before any appreciable sediment consolidation (fig. 3.4), judging from the nearly perfect state of many delicate-cell tissues. Carbonate and pyrite petrifactions are both relatively common and they indicate special marine sediment/water interface conditions which may be quite local; pyrite frequently occurs outside and investing the plant organ as well as in the cell cavities (see Hudson and Palframan 1969). In the case of some more robust woods the petrification could well have been rather later in diagenesis, the petrifying minerals (carbonate or silica) being derived from pyroclastics and other interbedded rocks.

Plant debris fossils which are visible in the field but normally only recognisable in the laboratory, and palynomorphs which are not visible in the field, are both almost always compressions (see fig. 3.3).

3.3 DISTRIBUTION OF FOSSILS

Land-plant fossils probably mostly represent lowland floras developed on aggradational land in deltas and only to a very small extent any other floras. These aggradational areas were and still are the most favourable for plant growth and thus probably have been the sites of the most intense biologic evolution by natural selection. This fortunate accident of preservation appears to provide the general evidence of land-plant evolution that is primarily required. In detail however, such plant megafossils probably represent more accurately the riverine and delta back-swamp elements, and this may imply some distortion of the full interpretation. Palynomorphs on the other hand, which seldom agree closely with the megafossils in identification from the same samples, probably record a much wider spectrum from the whole aggradational land, and sometimes from further.

In theory at least fossils of an upland flora could accumulate in high-level lake basins and in Tertiary times some genuine upland lake basin floras were preserved, e.g. in the Western United States; these deposits have not yet been recycled by erosion, but circumstances of such delay are thought to be both exceptional and geologically temporary. No Pre-Tertiary examples have been noted.

Frequently leaves, fruit, wood and pollen of one plant taxon are separately concentrated under control of (*a*) the sedimentation location for particles of a certain size together with plant fossils of that size, and (*b*) the rate of sinking of plant organs from direct waterlogging of their cell cavities. Thus, at any one locality, only one of these organs of any particular plant taxon is likely to be preserved. It also cannot be too strongly emphasised that although the term is often loosely used, a palynomorph assemblage is in no sense a 'microflora' nor can it represent transportation of the whole flora from any one land site; sedimentation and sinking rates, and distances from sites of the growing plants, are as effective controls below a size of 100 μm as they are with the megafossils (see fig. 3.5).

3.4 RESTRICTED AVAILABILITY OF CHARACTERS IN FOSSILS

In a living woody plant, the details of the following 18 items might each be considered at a selected level as a character: habit, *leaf morphology*, *leaf cuticle structure*, nodal anatomy, *wood anatomy*, inflorescence, flower perianth, stamens, carpels, ovule, *seed*, *fruit*, *pollen*, chromosome number, fertilisation, chemotaxonomy, cotyledons, root anatomy; these items could also either be subdivided or grouped because the selection of character units can only be an arbitrary process. A fossil plant of the same group, however, would probably be considered almost completely assembled if based only on the six characters shown in italics; none of the other characters are even remotely likely to be available. In most cases a 'fossil plant record' may consist of two organs assembled or even of only one. It is important therefore to acknowledge these discrepancies in information available, and to accept that all angiosperm phylogeny must be based solely on characters from the leaf, wood, fruit and pollen, with only very occasional additions of information from other organs (see fig. 3.6). Flowers are designed in life as ephemera, and their preservation potential is extremely low (see Leppik 1971). The data presented by Graham and Graham (1971) in a study of the Lythraceae indicate the nature of the problem well, although Eyde (1972) has pointed out many other shortcomings in their paper.

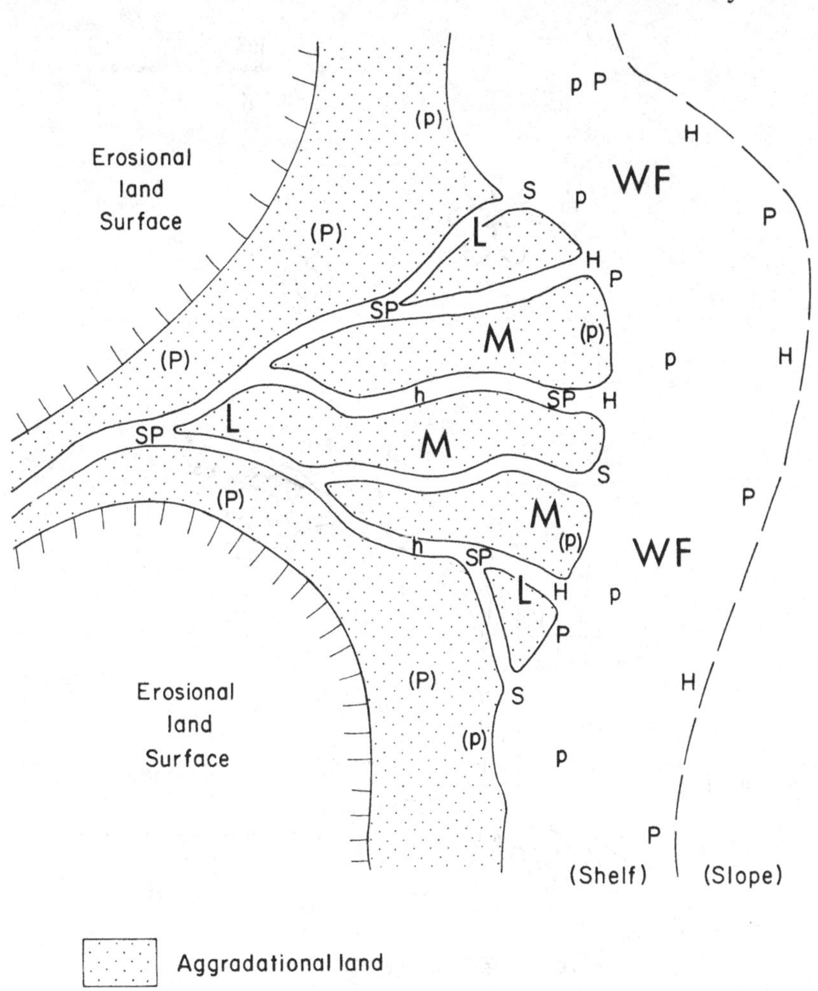

Figure 3.5. Diagram of plant fossil dispersal by water in aggradational land area and adjacent continental shelf; placing of individual letters indicates in each case an 'envelope' of maximum occurrence frequency to be expected. *L* = leaf; *W* = wood; *F* = fruits and seeds; *M* = mixed megafossils; *S* = spores; *P* = pollen; *p* = coastal plant pollen; *H* = marine plankton; *h* = tidal limit of plankton (restricted varieties). (*P*), (*p*) = wind distributed pollen.

The advent of scan microscopy and of micro-chemical investigations greatly improve the scope of palaeobotanic work, but they in no way change the discrepancy between Recent and fossil available characters nor affect the nature of the solution to the main problem under discussion.

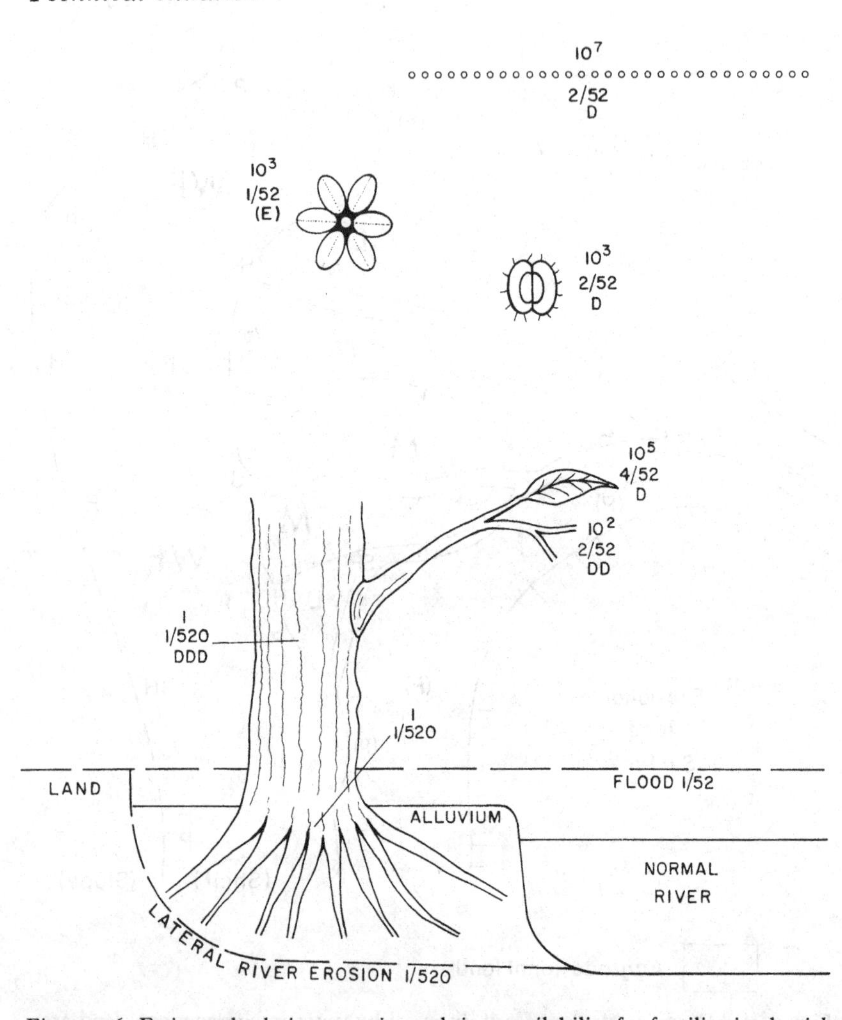

Figure 3.6. Estimated relative quantity and time availability for fossilisation burial of plant organs of arboreal angiosperm. 2/52 (for example) = two weeks in the year; flowers are ephemeral (E), fragile and lacking durable (D) tissues.

3.5 UNIQUE NATURE OF THE PROBLEM

There is perhaps no other group of organisms with so many living representatives and so little certainty about their phylogeny. The Insecta form a larger living group, with a longer geological history, with considerably less palaeontologic information and with very little known about phylogeny; the classification of their extant members however is relatively well established and little challenged, while in contrast in the angiosperms it is still necessary to state at all times which of several postulated hierarchical arrangements of taxa is being followed. It is possible that small animals are more easily viewed as entities and can thus be handled with more confidence.

Of other important living groups the evidence on fossil birds is altogether too fragmentary to consider, but fossil mammals have an appreciably later main radiation than angiosperms and have been handled with confidence so that their classification is well integrated with that of the Recent members. The fact that reconstruction and identification must so often have depended on the evidence of one bone or tooth, and yet have been possible, sharpens the idea that one taxon of organisms has but one composite character which can only be divided arbitrarily; the exercise of selecting or erecting unit characters for any organism past or present is merely part of the current human pattern of thought in this field.

One conclusion is that although this problem may be unique in its magnitude and difficulty, a new data-assembly and data-display method will ultimately be required.

4 Data-handling for fossils

The current system for handling all fossil data is what is loosely known as Linnean, which means that fossils are handled, as if they were living organisms, principally through the two still separate Codes of Botanical and of Zoological Nomenclature. The greater part of both these codes is concerned with very legalistic solutions to the continuing problems of typification, priority and synonymy. Although many palaeontologists seem curiously content with this arrangement, they are always greatly outnumbered in code discussions by botanists or by zoologists and have made little impact on the codes in terms of adequate provision for fossils. For example palaeobotanists have on more than one occasion just managed to avoid being forced to write their diagnoses in Latin simply because other botanists do; while palaeozoologists have completely failed to achieve recognition for parataxa in their code, although palaeobotanists have always used them.

I have suggested (Hughes 1971) that this arrangement is entirely inappropriate and inadequate as the only system for palaeontological material and that it is one of the major causes of lack of progress in the effective continued use of fossil data.

4.1 UNSUITABILITY OF THE CURRENT SYSTEM FOR PALAEONTOLOGY

The so-called 'Linnean' system is only a superficial grouping technique designed for a situation in which the definition of taxa can normally be tested by observation; perhaps however 'designed' is too strong a term for a system in which inertia is so important and a situation in which full user-comprehension is so exceptional that all control has passed by default from the user to the Code 'lawyers'. In detail the main problems appear to be:

(*a*) The only access to basic occurrence data is through the entirely uncontrolled subsequent author attributions to species

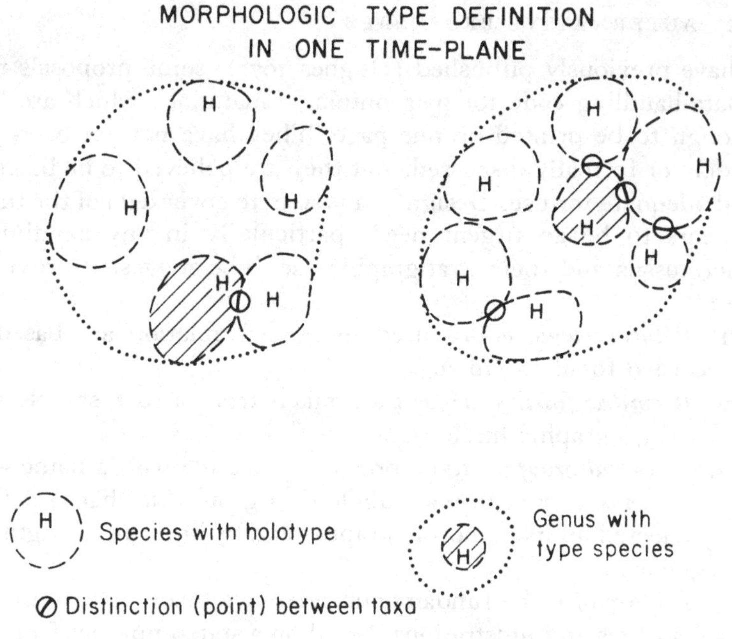

Figure 4.1. Standard taxonomic treatment of extant organisms with 'centre of cluster' type for reference. Distinctions between taxa may be given at certain points but are not obligatory.

and is subject in indexing to the vagaries of transfer between genera.

(*b*) The priority system for nomenclature makes the orderly addition of new data from the use of new techniques a problem rather than a natural and simple process.

(*c*) 'Synonymy represents one of the most intractable linkage-problems in palaeontology' (Mello 1969).

(*d*) The holotype (and type species) concept represents point-centred clusters (fig. 4.1) under the assumption that taxon limits are known from interbreeding barriers. With fossils the inter-taxon limits cannot be so known, have to be selected, and must be carefully defined; a point at 'centre' or elsewhere has no useful meaning, even if the type is said to have nomenclatural rather than taxonomic meaning.

By far the most important of these problems is the first because it actually restricts data-retrieval; the others merely cause waste of time, although as they belong to a carefully integrated system alternative proposals must cover them.

4.2 ALTERNATIVE PROPOSALS

I have previously published (Hughes 1971), some proposals for a 'Data-handling code for palaeontologic material', which are brief enough to be printed on one page. They have not yet been very widely or formally discussed, but they are believed to be balanced and adequate for use. Design of a system to cover data of the future appears to be an urgent need, particularly in any handling of microfossils and their stratigraphic use. The suggested provisions are:

1. *Palaeontologic events* used in time-correlation are based on recorded fossil specimens.
2. *Recorded fossil specimens* are taken from a rock sample with stated geographic limits.
3. A *formal group* of fossil organisms is a file with a name — .
4. A *genusbox* is a formal subfile of a group file that is defined by selected limits in morphography, and by limits in stratigraphic sequence.
5. A *biorecord*, the fundamental reference taxon, is expressed as a diagnosis and illustrations, based on a stated number from one sample, of recorded fossil specimens with normal distribution of variation of the characters studied. Nomenclature includes formal reference to group file and genusbox subfile, to date of record observation, author initials and serial number, and to a stratigraphic time-scale division.
6. All other recorded specimens are replaced in *comparison records*, which are formally compared with one or more biorecords on a graded scheme that expresses distance (see fig. 4.2).
7. All other terms are regarded as informal.

The following points are implicit in the use of this code: (*a*) retrieval of all record data may be made via rock description unit, or stratigraphic time-scale divisions, or geographic data or author or date or by a combination of these; (*b*) the principle and practice of binominal nomenclature are retained as the biorecord must be placed in a genusbox (fig. 4.3); (*c*) the biorecord depends only on the original material and description, with the result that it cannot be emended and it bears no priority; (*d*) new biorecords *may be* published without reference to earlier ones of similar age and scope, even in the extreme case using some or all of the same material; (*e*) comparison records must bear an opinion expressed as a grading (see Hughes 1973*c*).

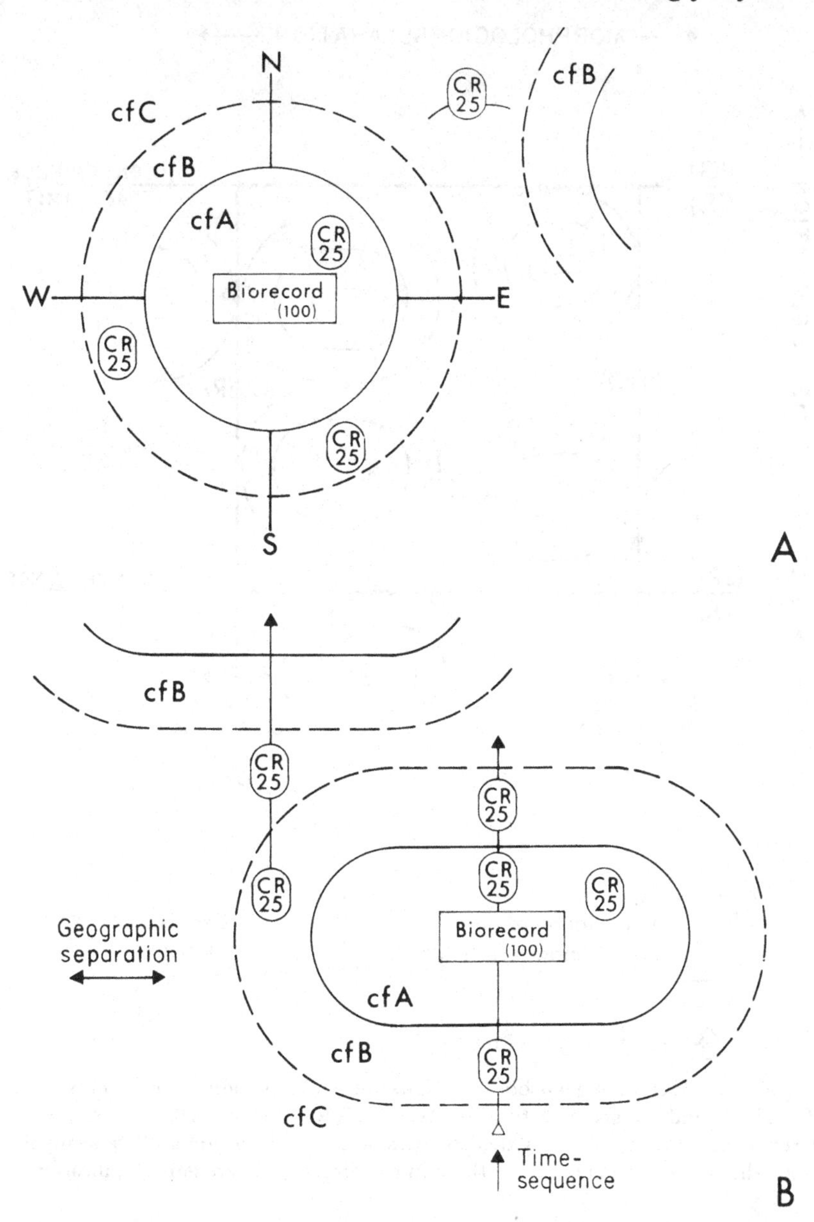

Figure 4.2. Diagrams of biorecords and graded comparison records. CR25, comparison record based on 25 specimens (in distinction from the biorecord based in this case on 100 specimens). cfA, cfB, cfC are successive grades of estimated comparison (defined in Hughes 1973*c*). A Morphologic fields of comparison records in space only. B Morphologic fields of comparison records in time on parts of two geologic sections overlapping in time. (After Hughes 1973*c*.)

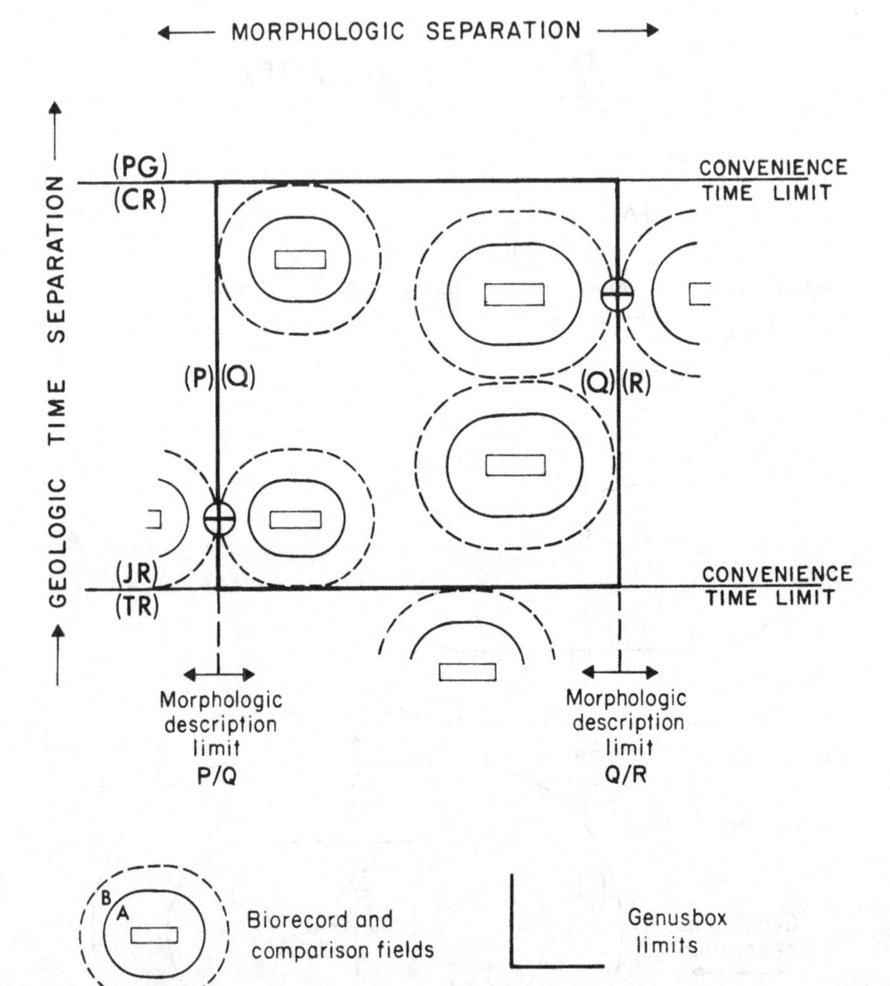

Figure 4.3. Erection of genusbox *Q* with its definition by morphologic limits from *P* and *R*, and by geologic time limits, e.g. TR, Triassic, JR, Jurassic, CR, Cretaceous, PG, Palaeogene. Other comparable morphologic limits will be available if another dimension is added to those in the diagram. There is no typification.

Because no types, no synonymy and no priority are involved, no other rules are necessary and the only additional details would be editorial concerning the mode of data storage. This code has been used in Hughes and Moody-Stuart (1969), in Hughes and Croxton (1973) and in a few other papers.

There is no reason why, if desired, the Linnean Code procedure should not be used for expression of biological opinion about fossils, in parallel with this new proposed code for recording and stratigraphic use of material. The two can be separate and complementary.

4.3 STRATIGRAPHIC TIME-CORRELATION

No real advancement of knowledge of early angiosperms or of any comparable problem will be made without improvements of stratigraphic time-correlation. At the simplest level time-correlation concerns the discussion of two floras (or faunas) and their comparison by such criteria as Simpson's (1960) resemblance index. Unfortunately it has long been the practice to search for equality between floras or faunas in different geological sections; this approach is in error because 'equality' is never possible. It is important on the other hand to interpret time-correlation as seeking a succession or ordering of events (fig. 4.4). The assemblage *X* in newly described section *B* should be time-correlated between two events on reference scale section *A*; as the number of events on reference scale section *A* increases or as knowledge advances the correlation bracket (e.g. event Q +[1970]− S event) can be narrowed by a new or refined statement (e.g. event T +[1975]− U event). The forcing of an 'approximate equality' precludes any future discussion of refinement (see Hughes and Moody-Stuart 1969, pp. 84–5).

At another level of correlation of multiple events in two rock successions Shaw (1964) has discussed such devices as regression statements for obtaining best fit. Such procedures are quite satisfactory for obtaining a single best answer up to a given moment of investigation. What is necessary however is the development of a new 'mathematical' rolling statement which is designed for continuous refinement. This has not yet been done, and would in any case be dependent on the logic of succession rather than on that of equality of individual events.

4.4 PALAEOPALYNOLOGY

Palaeopalynology differs from most other sectors of palaeontology in that fossils are normally present in such abundance that observation has to be very selective both of specimens and of characters because it would otherwise require unlimited time. This is the

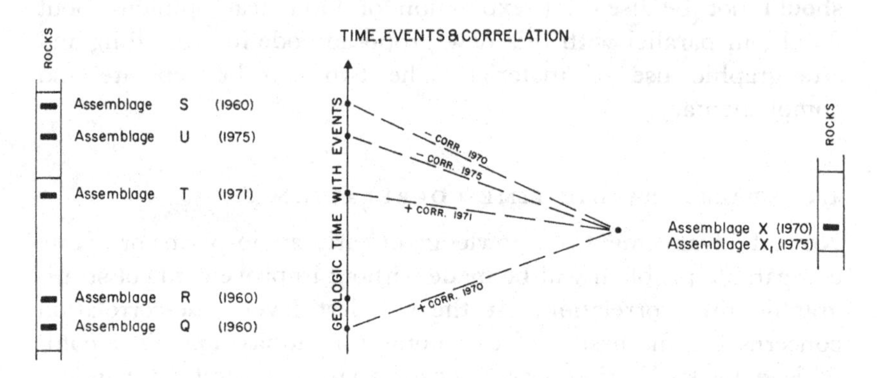

Figure 4.4. Time-correlation by event succession brackets. Given the knowledge available in 1970, assemblage *X* using existing methods might have been equated with scale assemblage *R* but no subsequent logical progress would have been possible. By the new suggested method, a new addition (*T*) to the reference scale in 1971, and the discovery of new microfossils in *X* giving a total assemblage X_1, both successively made possible a narrowing of the correlation bracket to event T +[1975] − U event; further discoveries could automatically lead to logical refinement of correlation in a similar manner.

opposite of the procedure in most megafossil palaeontology in which all conceivable characters are sought, and it requires the data-handling and stratigraphic techniques mentioned above in order to make proper use of this palynologic circumstance. Unfortunately much that is published in this field does not take this difference of abundance into account, and is in consequence relatively purposeless and of little stratigraphic value; most of such papers give considerable space to long-ranging taxa which cannot, by definition, contribute any stratigraphic information. The techniques required have as yet by no means reached perfection (see Hughes and Croxton 1973).

4.5 PALAEOBOTANY

Megafossil plants would naturally form the most convincing evidence of early angiosperms. Although they may occur in quantity at any one locality, the horizons concerned are always too few for megafossils to aid stratigraphic correlation significantly; even the circumscription of the megafossil taxa is influenced by the chance isolation of any one record surrounded by relative absence of other specimens in both space and time. The information from them, however, needs to be integrated with that from palynology; it should therefore be assembled and handled in the same way (e.g. Creber 1972). There is in any case no logical difference between types and sizes of fossils, and in palaeobotany there are also numerous occurrences of dispersed plant-fragment fossils (seeds, cuticle, wood fragments) as intermediates.

Within the existing handling system there is a serious difficulty in taxonomy and nomenclature of Tertiary plants, particularly from the Neogene. Many Palaeogene and all older plants are named as individual fossils even in the Linnean system, whereas virtually all Quaternary and many Neogene fossils are named as if they were parts of extant species. This latter is the Quaternary biologists' custom adopted on the ground that the time-span was too short to detect evolution; it negates, however, any attempt to use fossils in stratigraphy except in a non-evolutionary and somewhat debatable interpretative role using climates. Extension of this practice back into the Tertiary additionally involves quite unacceptable extrapolation. As palaeopalynology is also concerned it seems clear that use of a 'biorecord' system throughout would at least leave the data available for any subsequent interpretation.

4.6 PAST AND FUTURE

When any change of system is advocated, concern is expressed over past data in the literature. With a considerable amount of skilled work, almost any data can be disentangled from species names and re-recorded as comparisons; the problem will normally be to decide on a cost basis how much is worthy of such treatment.

For the future, the elucidations of the evolutionary history of angiosperms and their immediate ancestors will depend not on preferred data but on the integration and manipulation of all available information which must be neutrally recorded to make this integration possible.

PART 2

Cretaceous Earth history

5 Cretaceous floras

5.1 GENERAL STATE OF LAND-PLANT EVOLUTION

In spite of a time lapse of 50–60 Ma from mid-Jurassic to mid-Cretaceous, the early Cretaceous floras have seldom been clearly distinguished from those of early and mid-Jurassic because the proportions and nature of their two main megafossil constituents, gymnosperms and pteridophytes, were assumed not to have changed. There is little difficulty, however, in distinguishing Cretaceous palynologic assemblages, and on general grounds the often postulated long period of no floral change seems most unlikely.

Speculation about the number of early Cretaceous taxa relative to the present-day pattern, is perhaps only useful for the purpose of obtaining an order of magnitude, and the drawing up of a table such as 5.1 probably raises many further problems. It is important, however, to emphasise in doing so that the figures given relate in each case to the supposed number of taxa existing at an instant of time, and not to the number collectively present during the whole evolution of a geological period.

The number of Recent pteridophytes is dominated by ferns of the general group Dennstedtiaceae, both epiphytic and otherwise dependent on the great angiosperm diversity of the tropical lowlands; this expansion probably started in late Cretaceous time. Before that, in early Cretaceous time, Dipteridaceae and Gleicheniaceae had declined in numbers, but Schizaeaceae had increased giving perhaps a net overall small increase. Gymnosperms were clearly in greatest number in Jurassic and Cretaceous time, may have declined when so rapidly overtaken by angiosperms in late Cretaceous time, and then perhaps diversified slightly in their new high altitude and high latitude environments to what we now see. Is it possible to gauge how much more diverse mid-Mesozoic gymnosperms were than now? It is difficult to accept that they were more than three times as diverse, particularly as all such robust plants had a good chance of entering the fossil record. The end-Cretaceous angiosperm

Table 5.1. *Estimated number of seed-plant and pteridophyte species taxa existing at the selected instants of time*

Selected instant	Ma	Gymno-sperm	Pterido-phyte	Angio-sperm	Approxi-mate total
Recent	0	640	10000	286000	300000
End Cretaceous	65	500	2000	20000	22500
Beginning Cretaceous	135	1500	1500	0	3000
Mid Jurassic (end-Bajocian)	170	1500	1000	0	2500
Late Carboniferous (end-Westphalian)	300	200	300	0	500

Recent plant figures from Grant (1963); the estimated total of Recent non-vascular plants is about 100000.

estimate is based on evidence such as that of Chesters *et al.* (1967) *faute de mieux*.

5.2 HABITATS OF CRETACEOUS PLANTS

In forming a picture of Cretaceous vegetation, it is important to allow for a much lower degree of colonisation of possible habitats than obtains now. It is also difficult, when most living plants of the gymnosperm group now grow in quite other ways, to allow for gymnosperms (and of these, mainly coniferophytes) to have provided all tropical lowland vegetation; there could have been no other arborescent plants. It seems very probable that their precise life form is not yet understood. All fossils of trees are fragments; reconstruction is particularly difficult without petrification evidence, and it is very easy to use inadvertently a living plant model; such a model would be totally inappropriate for this problem.

If Cretaceous upland plants are among the known fossils, their reconstruction could legitimately be more like living gymosperms; however by all palaeobotanical experience and reasoning they will be rare and very fragmental. Thus any underestimate of the quantity or importance of upland vegetation scarcely affects the main interpretation problem.

Table 5.2. *Percentage of palynomorphs present in three adjacent lithologies in one section*

	'Lignite'	Shale	Limestone
Bisaccates	45	7	7
Classopollis	2	21	50
Gleicheniidites	3	8	2
Cyathidites	15	3	1
Cicatricosisporites	7	2	0
Others	28	59	40

The possible Cretaceous sea-margin mangrove habitat has been in dispute for some time. It has been suggested (Hughes 1973*a*, p. 190, and earlier) that *Classopollis* and its *Cheirolepidium/Brachyphyllum* type of foliage grew on the seaward margin of lowland forests and that such a distribution would explain the pollen occurrences. An example from the Aptian of France taken from Médus and Pons (1967, p. 111) is given in table 5.2. It seems most probable that *Classopollis* is high in the marine limestone and shale because it was directly carried there from a sea-margin nearby. The opposing argument is one of upland origin of these plants; when the sea rose and flooded the lowland forest it deposited shale and limestone adjacent to previous upland (Chaloner and Muir 1968). This argument really originates with discussion of Carboniferous *Endosporites* pollen/*Cordaites* plant in the same way (Neves 1958, Chaloner 1958), before Cridland (1964), using petrifactions, showed *Cordaites* to be a plant growing in water with stilt roots. In both the Carboniferous and Cretaceous cases the 'upland' theory comes from misunderstanding or 'over-dramatisation' of the expression 'marine transgression' (see Hughes 1975). Whether or not any real mangrove habit had evolved, is not yet known.

Finally it is strongly suspected that numerous water plants had developed in the Mesozoic, judging by the relatively large number of dispersed pteridophyte megaspore species known. Unfortunately water plants develop little cuticle on vegetative parts and so are unlikely to make durable compression fossils; in contrast their reproductive structures need unusually heavy protection from water and so make good fossils. In Mesozoic rocks such plants were clearly all of pteridophyte origin; by the above reasoning first evidence of angiosperm water plants could be expected from fruits, but not from leaves.

5.3 CRETACEOUS PLANT CHARACTERS

The most prominent general character of Cretaceous gymnosperms is the thick cuticle on leaves and seeds which is responsible for good preservation of compression megafossils, particularly in plant-fragment beds. Most pteridophytes do not share this character which was presumably xeromorphic and connected either with average height in life of the organ above ground level or more probably with long duration of function of the leaf (perhaps several seasons). The subsequent angiosperm development of expansion of the leaf lamina with a much thinner cuticle presumably represented a strongly contrasting life form. The frequently preserved flowers of the Benettitales were mostly large and heavy with a relatively thick cuticle, perhaps indicating longer time of function than the subsequent contrasting angiosperm flower which was delicate and ephemeral in purpose and as a result very seldom preserved at all; this delicacy could have been connected with changed co-ordination with insect life but as will be seen below there is no direct evidence of this.

Growth rings in secondary wood are often prominent, although their mere presence only indicates that the vegetation was non-equatorial, which applies to most floras. The pattern of growth within individual rings has so far been little investigated (see Chaloner and Creber 1973) but may be made to reveal more about regional climate. Modern techniques of detailed observation, of measurement, and of data-storage, may also yield information on astronomical data from the quality of groups of rings.

It is not yet possible to reconstruct life-forms (in the sense of Raunkiaer) from these fossil plants because so little information has been obtained from the Cretaceous about root structures. The removal from its growth position and the transport of the complete stem and root of any plant is unlikely. Roots alone are occasionally seen in soil beds but they are as yet unidentifiable without other organs; it also seems probable that in most cases they only represent *Equisetites* and a few other plants that grew in flooded or intermittently flooded sites.

Although the main Cretaceous ground cover can only have been of ferns or at least of pteridophytes, its distribution and detailed nature are not known in a way which might be possible if 'coal-ball' petrifications were available.

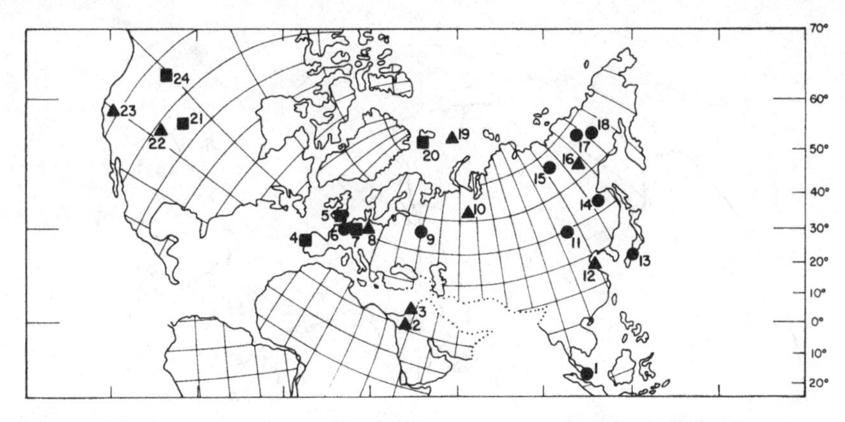

Figure 5.1. Early Cretaceous floras (approximately Berriasian–Barremian); northern Cretaceous hemisphere map (Smith *et al.* 1973). 1, Pahang, Malaya (Smiley 1970); 2, Jordan; 3, Lebanon; 4, Val de Lobos, Portugal (Teixeira 1948); 5, Wealden, England; 6, Belgium; 7, Buckeberg, West Germany; 8, Quedlinburg; 9, Moscow region; 10, West Siberia; 11, Gusino, Transbaikalia; 12, Shantung, China; 13, Honshu, Japan; 14, Bureja; 15, Lena-Vilui; 16, Aldan; 17, Indigirka; 18, Silyan–Kolyma–Zyrianka; 19, Franz Josef Land; 20, Vestspitsbergen; 21, Black Hills, N. Dakota; 22, Fuson, Wyoming; 23, Horsetown, N. California; 24, Kootenai, Montana–Alberta.

■ = Berriasian–Valanginian; ▲ = Hauterivian–Barremian; ● = Berriasian–Barremian, whole or part.

5.4 CRETACEOUS FLORAS

Despite great advances in palynology of dispersed microspores, the main megafossil floras will remain the key to the understanding of plant life of this period. Although these floras are as numerous as in other post-Silurian periods, it is important to recognise that in each case only a few horizons are rich in fossils and thus very short time-sectors are represented by the published 'flora'; the well-known British Wealden flora (Seward 1894–5), for example, comes mostly from a few horizons of Berriasian age and there is a very incomplete record from the remaining three-quarters of the time of deposition (Valanginian to Barremian) of the Wealden Group.

It is most important for this interpretation that not only should the stratigraphic position be quoted as accurately as possible but also the Cretaceous palaeolatitude of the occurrence; its present geographic position is immaterial. The main relevant Cretaceous floras are consequently plotted in three separate groups on the palaeolatitude maps of Smith, Briden and Drewry (1973); (*a*) Berriasian to Barremian (figs. 5.1, 5.2), (*b*) Aptian and Albian (figs. 5.3, 5.4), and

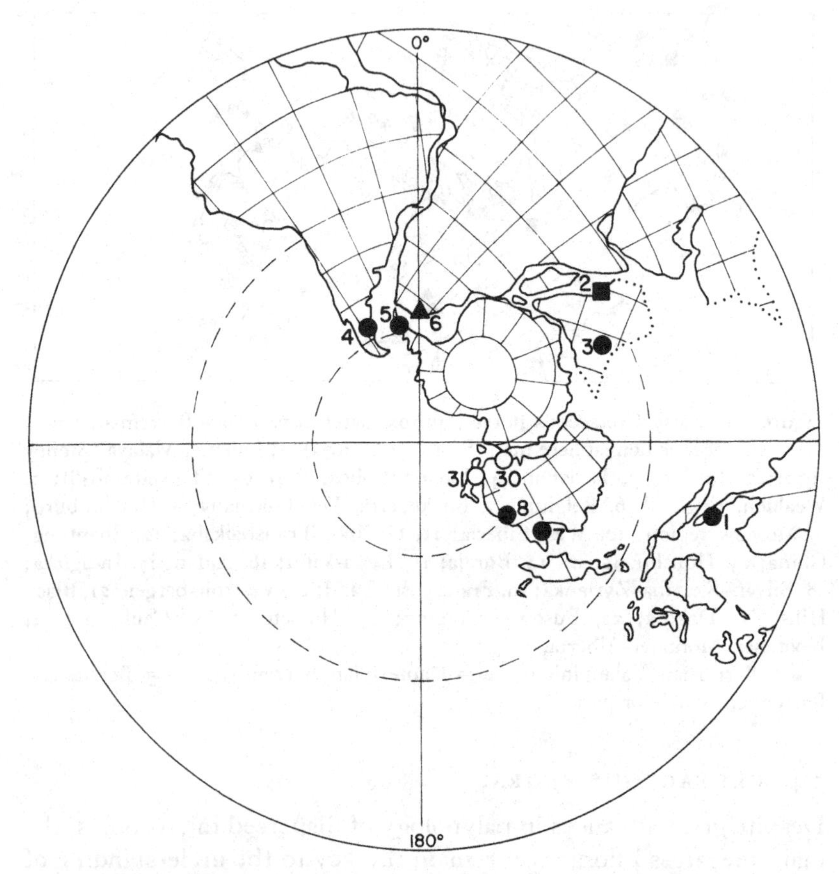

Figure 5.2. Early Cretaceous floras; southern Cretaceous hemisphere (polar projection) map (Smith *et al.* 1973). 1, Pahang, Malaya; 2, Cutch; 3, Jabalpur; 4, Santa Cruz, Argentina; 5, Grahamland; 6, Uitenhage, South Africa; 7, Plutoville, York Peninsula; 8, Burrum–Styx, Queensland. From Aptian–Albian (fig. 5.3): 30, Otway basin, Victoria; 31, South Island, New Zealand.
■ = Berriasian–Valanginian; ▲ = Hauterivian–Barremian; ● = Berriasian–Barremian, whole or part.

(*c*) Cenomanian and Turonian (fig. 5.5). These maps may subsequently be of use for plotting additional information and even for organised search among palynologic data. They also show more clearly the real nature of the gaps in knowledge and the value of the base from which extrapolated statements have to be made.

Palaeolatitudes are quoted as, for example, 35° KrN for mid-England. The maps were drawn for mid-Cretaceous time (approximately 100 Ma ago), and no attempt has been made to correct

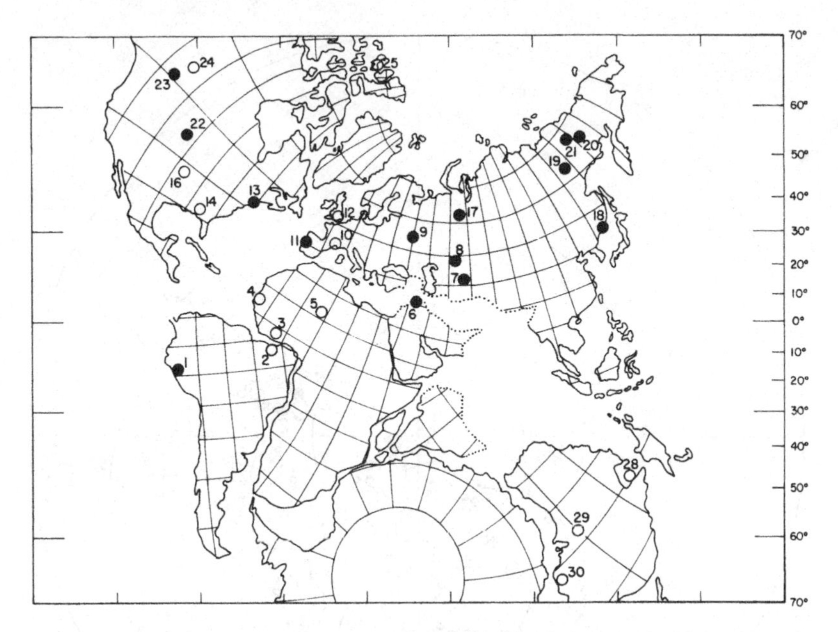

Figure 5.3. Mid-Cretaceous floras (Aptian–Albian); Cretaceous palaeomagnetic map (Smith *et al.* 1973). 1, Northern Peru (Brenner 1968); 2, Maranhao, Brazil (Herngreen 1973); 3, Côte d'Ivoire; 4, Senegal; 5, southern Tunisia (Reyre 1971); 6, Azerbaijan; 7, Kyzylkum; 8, Chushkakul, Kazakhstan; 9, Moscow region; 10, south France (Médus and Pons 1967); 11, Cercal–Buarcos–Nazare, Portugal; 12, southern England; 13, Potomac group; 14, Paluxy, Louisiana (Paden Phillips and Felix 1971); 15, Atlantic off Cuba (Habib 1968, 1969); 16, Woodbine, Oklahoma (Hedlund and Norris 1968); 17, west Siberia; 18, Suchan; 19, Aldan; 20, Silyan–Kolyma–Zyrianka; 21, Indigirka; 22, Lakota, S. Dakota; 23, Blairmore, Alberta; 24, Mannville, central Alberta (Singh 1964); 25, Christopher formation, Canadian Arctic Islands; 26, Chandler River, north Alaska; 27, Corwin, north Alaska (Smiley 1969*b*); 28, Mornington Island; 29, Oodnadatta; 30, Otway Basin (Dettmann 1973); 31, Paparoa, Nelson, New Zealand (Couper 1953).
• = megafossils with or without palynomorphs; o = palynomorphs only.

further for different parts of Cretaceous time by interpolation between the mid-Cretaceous map and maps for mid-Jurassic or Eocene time. The rates of plate formation in different areas and at different times vary considerably, and thus also the rotation and movement of the continental plates; it is better therefore to wait for completion of the planned computation of these intermediate positions. In nomenclature 'KrN' is adequate for the somewhat crude palaeolatitudes here given, but it can be suitably superseded by such abbreviations as 'AptN' (for Cretaceous Aptian) when more precise computed positions become available.

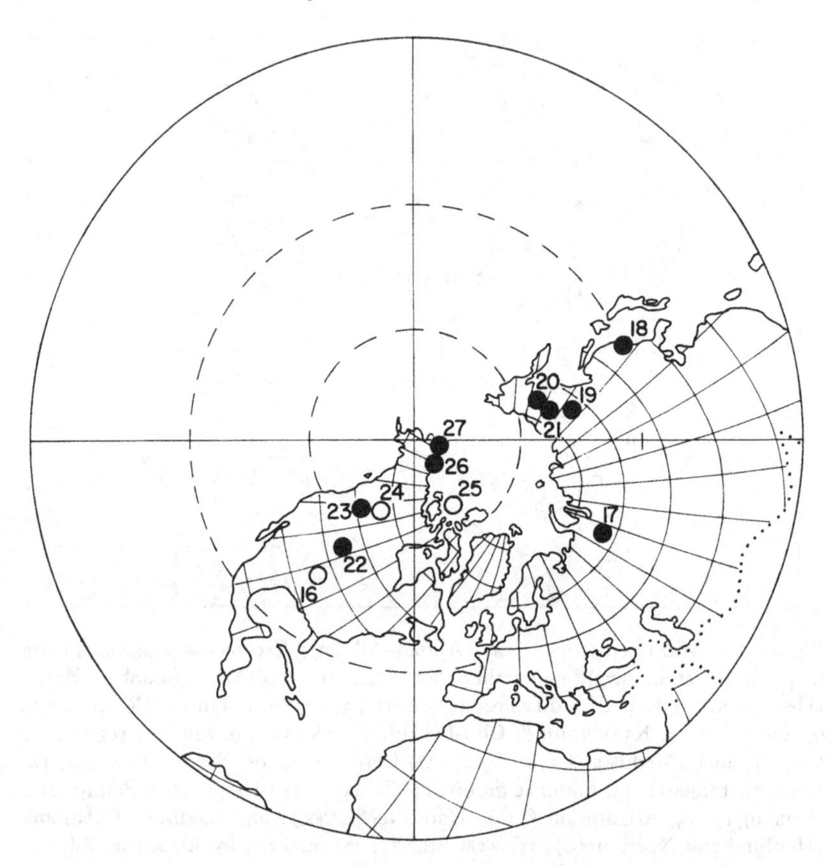

Figure 5.4. Mid-Cretaceous floras (Aptian–Albian); north polar palaeomagnetic map (Smith *et al.* 1973). Numbers and symbols as in fig. 5.3.

The palaeolatitudes of most of the principal European and North American mid-Cretaceous floras are plotted together with some assemblage details (fig. 5.6); from this diagram it is clear that no well-known flora falls between the Cretaceous equator and 25° KrN. In the latitudes up to 25° KrS (fig. 5.7) there is so far only an Albian–Cenomanian flora at 15° KrS (7° S, now) from Peru (Brenner 1968), and another at 15° KrS from the 'Neocomian' of Pahang, Malaya (Smiley 1970). The mid-Cretaceous palaeolatitude map (fig. 5.3) makes it clear that only the East Indies, southern China, Arabia, North Africa and the northern quarter of South America are available for search for Cretaceous tropical floras. Palynologic records (see Reyre 1970; Hughes 1973*a*, text-fig. 2) indicate abundance of *Classopollis* in those palaeolatitudes, falling off steadily

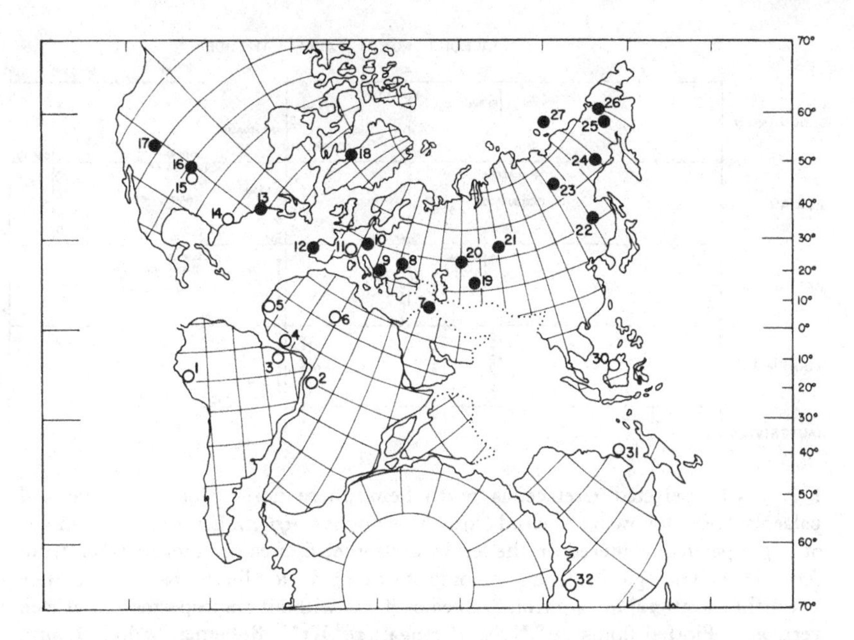

Figure 5.5. Cenomanian–Turonian floras; Cretaceous palaeomagnetic map (Smith *et al.* 1973). 1, northern Peru; 2, Gabon (Boltenhagen 1967); 3, Maranhao, Brazil; 4, Côte d'Ivoire; 5, Senegal; 6, south Tunisia; 7, Azerbaijan; 8, Kanev, Ukraine; 9, Bulgaria; 10, Peruc, Bohemia; 11, Rhône; 12, Tavarede–Bussaco, Portugal; 13, Raritan, New Jersey; 14, Tuscaloosa, Carolina; 15, Woodbine, Oklahoma; 16, Cheyenne, Kansas; 17, Dakota formation, south Utah; 18, West Greenland (see Koch 1964); 19, Kyzylkum; 20, southern Urals; 21, Yenesei; 22, Zesho–Bureya; 23, Lena–Vilui; 24, Archajel–Kolyma; 25, Grebenka; 26, Anadyr; 27, New Siberian Islands; 28, Corwin, Alaska; 29, Colville River, Alaska; 30, Sarawak (Muller 1968); 31, Bathurst–Melville Island (Dettmann 1973); 32, Otway basin, Victoria; 33, Mata, South Island, New Zealand (Couper 1953). Localities 28, 29 and 33 cannot be plotted on this map projection.

north and south, but the provenance pattern of this most interesting pollen has not yet been fully explained.

All the classic European and east North American floras fall in latitudes 25–40° KrN, but in higher latitudes up to the Cretaceous North pole in Alaska (Smiley 1972) there are numerous floras from Barremian age onwards. The southern hemisphere has floras in each continent but a less continuous record. In detailed knowledge all Cretaceous floras lag far behind those of the Carboniferous because so very little other than a few individual *Tempskya* and *Cycadeoidea* specimens has been discovered petrified.

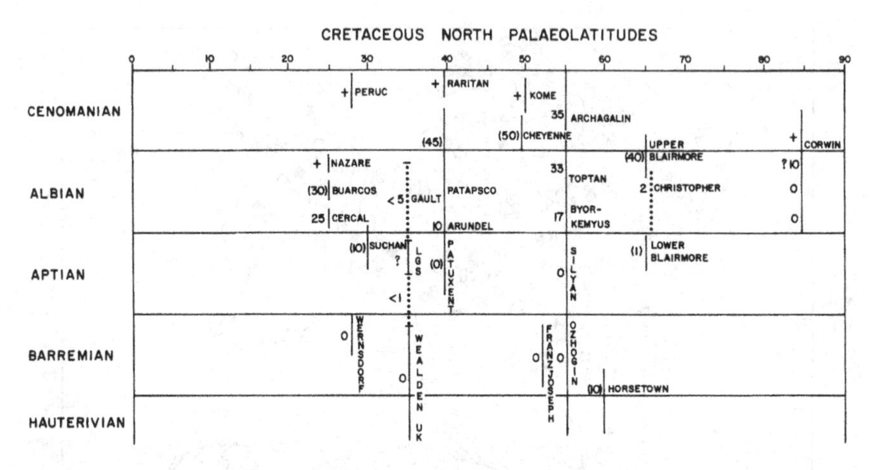

Figure 5.6. Selected Cretaceous north hemisphere floras plotted by age and palaeolatitude, following Axelrod (1959) and Hughes (1973*a*). Estimated percentage of angiosperms is entered on the left of each line; figures in brackets taken from Axelrod. Palynological information only shown by dotted line; cross-line on range = reliable stratigraphic control. Names of floras without angiosperms are written vertically. Plotted floras: 25° KrN, Portugal; 28° KrN, Bohemia, and Germany; 30° KrN, Vladivostok; 35° KrN, England; 39° KrN, Potomac Group, Maryland; 50° KrN, West Greenland, and Kansas; 52° KrN, Franz Josef Land, Arctic; 55° KrN, USSR north-east Siberia; 60° KrN, north California; 65° KrN, Alberta, Canada, and Canadian Arctic Islands; 85° KrN, North Alaska. LGS = Lower Greensand.

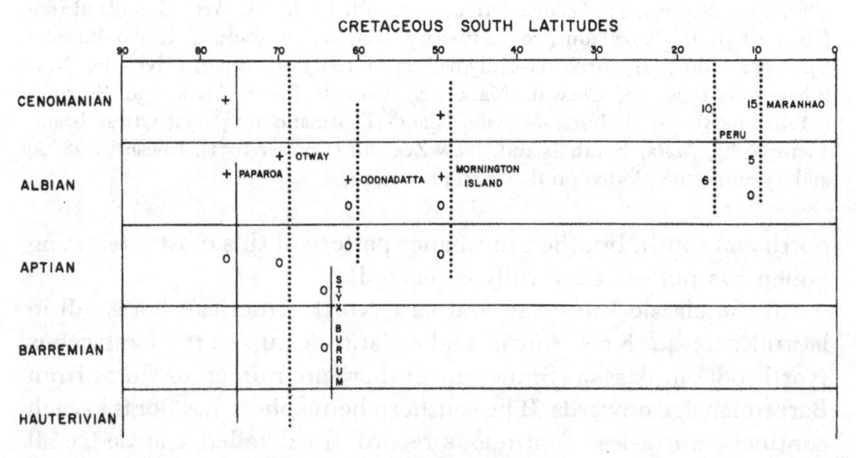

Figure 5.7. Selected Cretaceous south hemisphere floras plotted by age and palaeolatitude. Explanation as in figure 5.6. Plotted floras: 8° KrS, north-east Brazil; 15° KrS, north-east Peru; 48° KrS, north Queensland; 60° KrS, central Australia; 63° KrS, south-east Queensland; 68° KrS, Victoria; 75° KrS, South Island, New Zealand.

5.5 CRETACEOUS FLORAL PROVINCES

The most important attempt to recognise provinces has been by Vakhrameev (1964, 1970, 1971) in Eurasia, one of the only favourable areas for the purpose at this time. He refers for Early Cretaceous time to an Indo-European province and to a Siberian province with some subdivision. When replotted on maps of palaeomagnetic origin (e.g. Smith *et al.* 1973), the province boundary appears to be fairly close to a line of palaeolatitude (35° KrN); on palaeomagnetic grounds also, any reference to India in the northern hemisphere before the Eocene is better avoided.

The basis for this division is megafossils alone and, as can be seen from the relatively small number of plotted floras in such a large area as Asia (Vakhrameev *et al.* 1970, p. 261), the control is limited; in addition all Neocomian (presumed to mean Berriasian to Barremian) records are plotted together. Palynology supports these findings in a general way, e.g. the plot of *Classopollis* which falls to negligible proportions also at the palaeolatitude 40° KrN (Hughes 1973*a*, text-fig. 2).

The Siberian province floras have abundant ginkgophytes, Czekanowskiales, *Pityophyllum* and *Podozamites*, and *Coniopteris* spp.; there is no *Tempskya*, *Cycadeoidea* or *Williamsonia*. Coal formation was common, but the variety of arborescent plants was always less than that of the 'herbs and shrubs'.

In late Cretaceous time (Vakhrameev *et al.* 1970, p. 283) the equivalent province boundary was also near 40° KrN palaeolatitude but the eastern part of the Asian tectonic plate was much further north than before so that at this time most of present Siberia fell into the Siberian province.

Although it is not the intention here to analyse the details of these floral provinces, it is interesting to note that in mid-Cretaceous time Alaska was polar and most of western Canada in very high latitudes (Calgary 65° KrN), appreciably further north than the main Siberian province at that time.

The precise palaeolatitudes may be changed with advancing knowledge, although for post-Triassic time such changes are not likely to be great. The hope is that now the relative positions of Cretaceous floras are known, much more can be derived from study of the plants themselves without at the same time trying to estimate their geographic position from their nature; the circular argument can be avoided.

Figure 5.8. Reconstruction of early Cretaceous Wealden flora of southern England. (Reproduced with permission of the Institute of Geological Sciences.)

5.6 RECONSTRUCTIONS OF PLANT LIFE HABIT

It is very difficult to produce a satisfactory image of past life which does not at the same time impede further investigations. It has been found that pictorial reconstructions, particularly if they are skilfully arranged and drawn, tend either to suggest completeness of information or at best simply to guide the imagination to problems secondary to the picture; there is also a tendency for all details not carefully considered by the palaeontologist or by the artist to be 'uniformitarian' in concept, which inevitably means that they are wrong or misleading because evolution has been omitted. An ideal reconstruction should indicate the unknown areas as well as the ascertained facts, but it is difficult to make such presentations both attractive and generally educational, and few attempts have been made.

The best available early Cretaceous reconstruction (fig. 5.8) is that prepared and published by the United Kingdom Institute of Geological Sciences (Edmunds 1935, Gallois 1965) for the Wealden. It was clearly drawn with good co-operation between palaeobotanist and artist, but it illustrates well the several fundamental difficulties facing any such attempt, whether forty years ago or now. The restoration was made as much for the vertebrate animals as for the plants; this probably dictated the choice of an open shore scene rather than a delta riverine margin, back-swamp area or outer-face margin to the 'lake'. Such delta scenes would inevitably have been more confused and difficult to portray with clarity. The restorations of the individual *Equisetites*, ground-ferns, *Tempskya*, and cycadophytes, provide reasonable impressions, although in some cases subsequent publications have clarified both details and scale. The predominant vegetation, however, should have been the coniferophytes in the shape of the prevailing brachyphylls and linearphylls rather than of living *Pinus*, *Sequoia* and *Araucaria* trees. There probably were a few major straight tree trunks as depicted, because some fairly large logs have been found in contemporaneous deposits elsewhere, but there was certainly also much universal low-tree and scrub vegetation. It may appear to be impossible ever to reconstruct the form of such plants accurately from the fragments available, but if such details as the pattern of branching, the size of leaves, and the frequency and placing of cones, were carefully collated from a large number of specimens a new restoration technique could emerge. The general impression of this flora as being without angiosperms is accurately conveyed, for although the author (Edmunds 1935, p. 21) mentioned rare angiosperms none have specifically entered the reconstruction.

5.7 GENERAL QUESTIONS

Axelrod (1959) postulated equatorial or low latitude origin of angiosperms and their subsequent poleward 'migration'. The suggestion appeared on general grounds to be a good one although it is questionable whether the expression 'migration' is really suitable for plants, which simply disperse into and thus colonise a new area and so extend, or fail to do so. The perennial question of all palaeontology is whether the duration of migration of animals or dispersal of plants is ever long enough to exceed the finest recognised subdivision of geologic time in the period concerned. Such fine stratigraphic subdivision has by no means been achieved yet in the Cretaceous.

The data from Axelrod (1959) were replotted with some additions (Hughes 1973*a*) with palaeolatitudes rather than present-day latitudes; an updated version is given in figs. 5.6 and 5.7 and the conclusion from it is that stratigraphic correlation of non-marine strata distant from the marine reference scale in western Europe is not yet good enough to show any latitudinal gradient of arrival of angiosperms as originally suggested. Detection of such a dispersal gradient of these or of any other plants is, however, a worthwhile aim for palynologic as well as stratigraphic effort.

If the time from the Devonian period to the present day has recorded progressive degrees of colonisation of the land by plants, it follows that many possible habitats now filled would not have been filled in early Cretaceous time. The integration with land animal life was on the same reasoning also far from complete, although this will be discussed in the next chapter.

The general diversity of plants at the present time is so great that it is somewhat difficult to picture the evolving Cretaceous world with much less diversity both locally and over the latitude range. This would result in a weaker contrast between northern and southern hemisphere land plants, as is observed in the Mesozoic, although the principle of an equatorial belt of separation must have operated to some extent. Perhaps if the equatorial belt was as at least 80° of latitude wide, as is suggested by the distribution of *Classopollis*, there would be more appearance of uniformity. This relatively low diversity does not in any way imply any shortage of numbers of individual plants, if the general frequency of fossil remains in sediments may be taken as a guide. If such low diversity existed over great plant-covered areas, a kind of simple stability of the genetic pool may have been linked with it; this is another way of suggesting that polyploidy and many other genetic elaborations now observed may have mainly arisen with the very great expansion of angiosperms and of insects in Tertiary time.

The general small size of leaf in the conifers is puzzling because these plants must have been the dominants, at least in lowland areas, both in quantity and in individual size, judging from such petrified tree trunks as are occasionally available. Leaf size had already been much larger in Carboniferous *Cordaites* although probably this was development of a drip-tip style in an equatorial rain-forest environment. Leaf size in cycadophytes was greater but most evidence suggests that these plants were never very large or high-growing. There appears to be some physical evolutionary factor not yet understood, which makes satisfactory interpretation of this vegetation very difficult.

There seems little doubt that the succession of gymnosperm to angiosperm habit took place first in woody plants leaving angiosperm herbs as a subsequent development in relatively adverse higher latitude areas. The understanding of the general questions about early Cretaceous trees is therefore an important matter.

6 Cretaceous land fauna

The inter-relationship of plants and animals on land may be regarded as lying on a scale of development from rudimentary in early Devonian time to the complex ecological circumstances of today (Hughes and Smart 1967, Smart and Hughes 1972). The early Cretaceous (from 135 to 100 Ma ago) occupies a position on such a scale, and the problem is to understand the degree of integration by that time (fig. 6.1); it seems probable that there was a considerable 'vacuum' in many fields in which available plant facilities had not yet been used or developed by animals. The insects must have been the most advanced animals in this respect in Cretaceous time, and several of the many other animal groups now involved had not then made any such progress. These other groups will be discussed briefly before returning to the insects.

6.1 REPTILES

Reptiles clearly provided the largest individuals on the Mesozoic land areas, and also a great variety in small sizes. Although the habit of using plant vegetative organs as food was apparently very little developed before Triassic time, by Cretaceous time it was the mode of life of most of the large dinosaurs. This feeding was probably relatively indiscriminate, and not therefore directly connected with the nature of plant reproductive structures. Most of the smaller reptile groups had originally consisted of minor carnivores, with the base of the food chain concerned being either aquatic or land-plant trash at least partly bacterially decayed. Larger carnivores followed the fortunes of larger herbivores. Any reptile concerned with direct use of plant reproductive structures (and thus with pollenation and distribution) was probably of small size. However, low metabolic rate was characteristic of the reptiles, and connected with this was the tendency to giantism that was developed to its limits in the mid-Mesozoic.

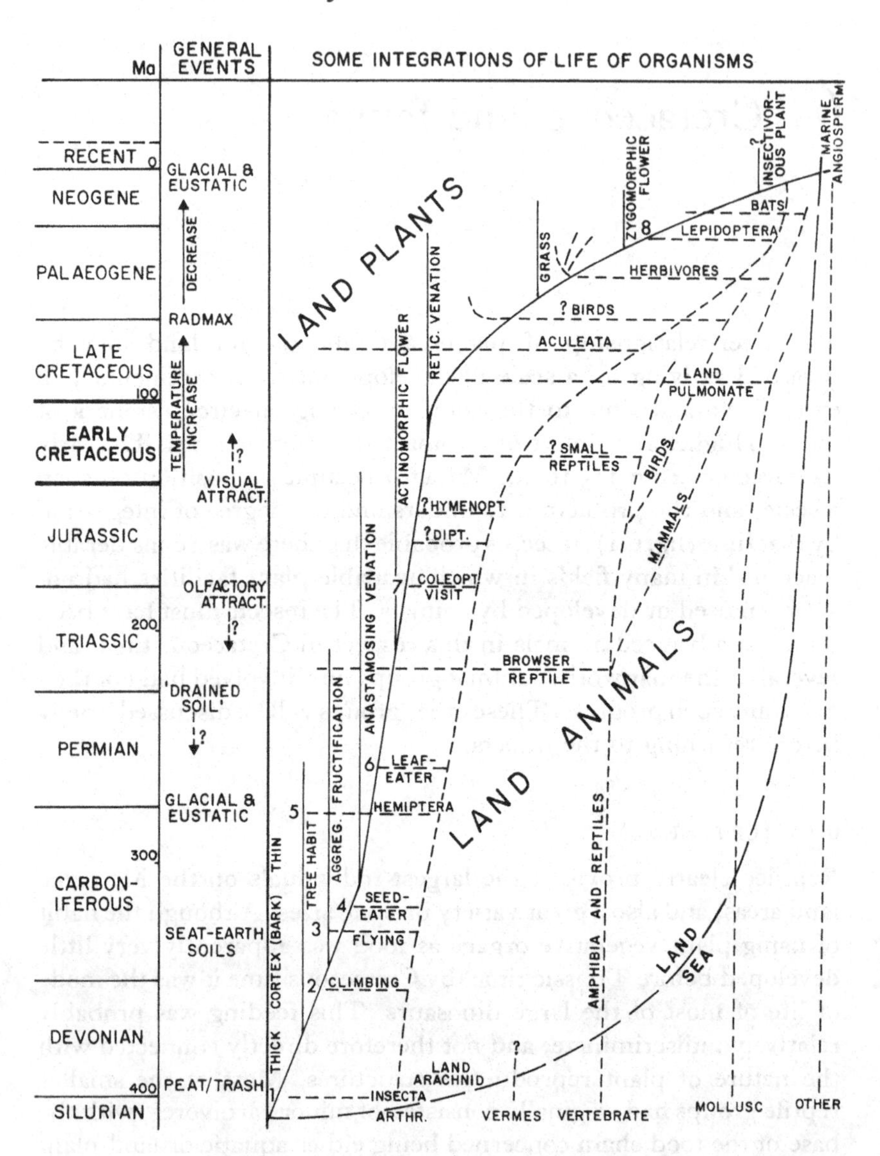

Figure 6.1. Experimental diagram to indicate some aspects of the progressive degree of integration of higher land-plant and animal life, through geologic time. The selected animal indicator lines refer only to possible integration with plant life, and have been drawn horizontally to avoid any evolutionary implications. Numbers (in the plant sector): 1, trash-eating; 2, tree-climbing flightless insects(?); 3, first fossils of flying insects; 4, fructification and seed-eating, inferred from detail of plant morphology; 5, plant-lice (Hemiptera); 6, leaf-eating, fossil evidence; 7, Coleoptera visit Benettite flowers(?); 8, lepidopteran, bat and bird flowers.

In detail the herbivorous reptiles of the early Cretaceous included the following: (*a*) sauropods (Saurischia), such as *Brontosaurus*, which were large and quadrupedal semi-aquatics, with small jaws and relatively weak teeth which suggest aquatic vegetation as food; (*b*) ornithopods (Ornithischia), such as *Iguanodon*, which were large bipedal forms with an efficient chopping dentition; *Hypsilophodon* was a small light form believed to have been cursorial (Galton 1974) but perhaps also a tree-climber; (*c*) Ankylosauria (Ornithischia) which were quadripedal and armoured with thick bony plates, but possessed small teeth. The well-known Ceratopsia (Ornithischia) or horned dinosaurs were very diverse but only evolved in late Cretaceous time at the same time as the angiosperms. Turtles (Anapsida) were present throughout but were of course slow moving ground herbivores.

Thus of the herbivores only the smaller ornithopods such as *Hypsilophodon* were likely to have lived on gymnosperm fructifications. Small theropod (Saurischia) carnivores such as *Ornitholestes* and perhaps some lizards (Squamata) probably moved about freely in the vegetation hunting insects etc. and could have been concerned in pollenation. Otherwise the large early Cretaceous reptile fauna may have sheltered in the vegetation but was unlikely to have been further integrated ecologically with the plant life.

6.2 OTHER VERTEBRATES

Among the few groups of small mammals in the early Cretaceous were the multituberculates, the first mammal herbivores; they seem to have been comparable with present-day small rodents but were probably relatively uncommon. Bats (Chiroptera) only appeared in the Tertiary.

Pterosaurs were apparently specialised aerial carnivores of fish, and must even have avoided fixed vegetation. Birds present some difficulty because of the improbability of preservation in most cases. The known record begins with late Jurassic *Archaeopteryx* which was toothed and presumably carnivorous but could well have been a woodland bird. The few early Cretaceous fossils are of water birds or waders; the late Cretaceous fossils are of the same kind although considerable radiation had occurred (Fisher 1967). The main Passeriformes radiation appears to have been in Palaeogene time or Maestrichtian at the earliest. Although it is possible that there were as yet unrecorded woodland birds in early Cretaceous time, there is no reason to suspect that seed-eating was an important bird development before Tertiary time.

6.3 INSECTS

The preservation record of all insects is poor and it so happens that there are very few useful records from the Cretaceous. In the English Wealden fossil elytra (wing-cases) of Coleoptera have been mentioned as cupedid beetles without description by Crowson (1962); this was from the Wadhurst Clay. From several lower horizons of the Hastings Beds, mainly Ashdown Sands, Binfield and Binfield (1854) and Westwood (1854) mentioned a considerable number of similar fossils and fragments. Fossils of this kind are probably more widespread than is suggested by their rare mention in the literature, and although it may now be possible to extract more information from them this has not yet been done. Reliance has to be placed on the fortunate sedimentary accidents which produce more complete faunas; Brodie (1845) described and figured such a fauna from the 'Wealden', but in modern terminology the beds concerned are the late Jurassic Purbeck beds of Wiltshire. Brodie (1854) described other Purbeck fossils from the Insect Bed of Durlston Bay, Dorset; even these, however, are probably not representative, as noted by Westwood (1854) who commented on the small average size of the specimens, a factor which presumably reflected conditions of sedimentation.

It is therefore necessary to consider here what was present in the late Jurassic of other areas and in the Palaeogene, in order to bridge this scarcity of fossils in Cretaceous time (fig. 6.2) through which at least seventeen orders of insects appear to have ranged (Crowson *et al.* 1967). Of the ten orders of Hemimetabola, most were carnivores (e.g. Odonata, dragonflies) or trash-feeders (e.g. Blattodea, cockroaches); only the Hemiptera (plant-lice) were intimately connected with living plants, although not in a way likely to effect pollenation or seed dispersal. In the Holometabola, only three orders, Coleoptera (beetles), Diptera (flies) and Hymenoptera (ants, bees, wasps), could be concerned; Neuroptera (lace-wings) were carnivorous.

Some extensive insect faunas are known from the late Jurassic, including those from Solnhofen (south Germany) and from Karatau in Kazakhstan, both of Kimmeridgian age. A Karatau fauna, recently amplified (Rohdendorf 1968), gives some idea of the composition of one of these assemblages totalling 18000 specimens, see table 6.1. Allowing for representation failure and for preservation failure this probably provides a fairly accurate view of the available insect groups in early Cretaceous time. The Lepidoptera, in which group the imago forms very close association with various angiosperm flower types, is first recorded from the Tertiary, and there seems every reason to believe that it evolved then with the angiosperms and was not represented in Cretaceous time.

Table 6.1

Family	Species	% of total specimens	Approx. average no. specimens per taxon
		Hemimetabola	
Ephemeroptera	1	0.1	18
Odonata	29	0.6	4
Blattodea	10	9.7	180
Orthoptera	5	2.2	70
Hemiptera[a]	27	9.9	67
Others	7	1.2	—
		Holometabola	
Coleoptera[a]	61	55.5	167
Neuroptera	12	1.8	25
Diptera	65	13.9	40
Hymenoptera[a]	54	1.8	6
Others	17	3.3	—

[a] Orders considered in more detail in the text below.

Behind this however, the main preservation difficulty is probably because in the Holometabola the larval stage, which is the chief feeding stage in most cases and could be considered to be the effective organism, is destined either to be consumed by predators or to grow into a quite different adult (imago); preservation of the larva as a fossil is an unlikely fate (Smart and Hughes 1972). In contrast, the imago or reproductive stage is destined either to be consumed or to become a fossil; this relatively ephemeral stage of the organism is the one usually identified and the only one to be identified fossil. The pupal stage in the Holometabola could by this argument be expected in theory as a fossil but would have neither characters of larva nor imago and might well be difficult to recognise.

	LATE JURASSIC	CRETACEOUS		PALAEOGENE
	SOLNHOFEN KARATAU PURBECK	WEALDEN (UK)	ALASKA	
Odonata	—			—
Ephemeroptera	—			—
Orthoptera	—			—
Hemiptera	—			—
Coleoptera	—	—	—	—
Diptera	—			—
Neuroptera	—			—
Hymenoptera: Symphyta	—			—
Aculeata			—	—
Lepidoptera				—

Figure 6.2. Table to illustrate the considerable information failure in most of the record of fossil insects of relevant groups in the Cretaceous period.

6.3.1 *Hemiptera*

The earliest records are from the mid-Permian for both Homoptera and Heteroptera (Wootton in Crowson *et al.* 1967), and they are strongly represented in the Jurassic. The host plants are not known, and although in the Tertiary the presence of certain plants may be inferred from the recorded Hemiptera this is not possible for Mesozoic fossils, and the chance of being able to do so seems remote.

6.3.2 *Coleoptera*

In the Karatau assemblage description referred to in table 6.1, the majority of the taxa recorded have been attributed to the Cupedidae, the Staphylinidae and the Chrysomelidae, although almost all in new genera and species. Crowson *et al.* (1967) were clearly doubtful about ascribing any of the previously discussed Karatau material to these extant families and pointed to various differences; unfortunately the new assemblage descriptions above (Rohdendorf 1968) was made just too early for any reference to be made to Crowson's views. Although there is no intention of analysing such matters here, it appears to be another case of the eager misuse of extant taxa as has occurred in the angiosperms and is discussed more fully elsewhere in this book. Cutbill and Funnell (1967, fig. 10, p. 807) illustrate the crude nature of this taxonomy together with the apparently almost complete Cretaceous failure of records (slightly modified by inclusion of some Alaskan discoveries from Langenheim *et al.* 1960, Langenheim 1964, 1969).

There were clearly a great variety of beetles (fig. 6.3) which could have moved about freely over all vegetation. It is probably more profitable to discuss the general effect of this than to pursue taxonomic affiliations. At the present day 'beetle flowers' are usually green or white, open, and odoriferous; they are characteristic of tropical areas and are rare in temperate Europe. Although beetle visits to flowers may appear to be accidental, they have been shown in many cases not to be (Faegri and van der Pijl 1966); most beetle mouth parts are orthognathic (perpendicular to the body axis) and thus well suited to chewing pollen but not to obtaining nectar except in a very open flower. Angiospermy and epigyny appear to be well arranged to protect ovules from these relatively clumsy 'mess and soil' insects which in a rough way may effect pollination. Recent cycad inflorescences also seem to attract beetles by the odour of some decaying protein. The robust Benettite flowers and Mesozoic *Nilssonia* group inflorescences appear certain to have been used in this way.

6.3.3 *Diptera*

This group has previously been recorded from the late Triassic and the Jurassic, and was undoubtedly represented in early Cretaceous time (Hennig 1970). At the present day, some small flies are associated with white and yellow shallow open flowers with nectaries, but as a group Diptera display very varied approaches to plants from 'mess and soil' to sapromyophily (Proctor and Yeo 1973); in general pollenation is effected irregularly and is thus unreliable. Perhaps more numerous flies are concerned with verte-

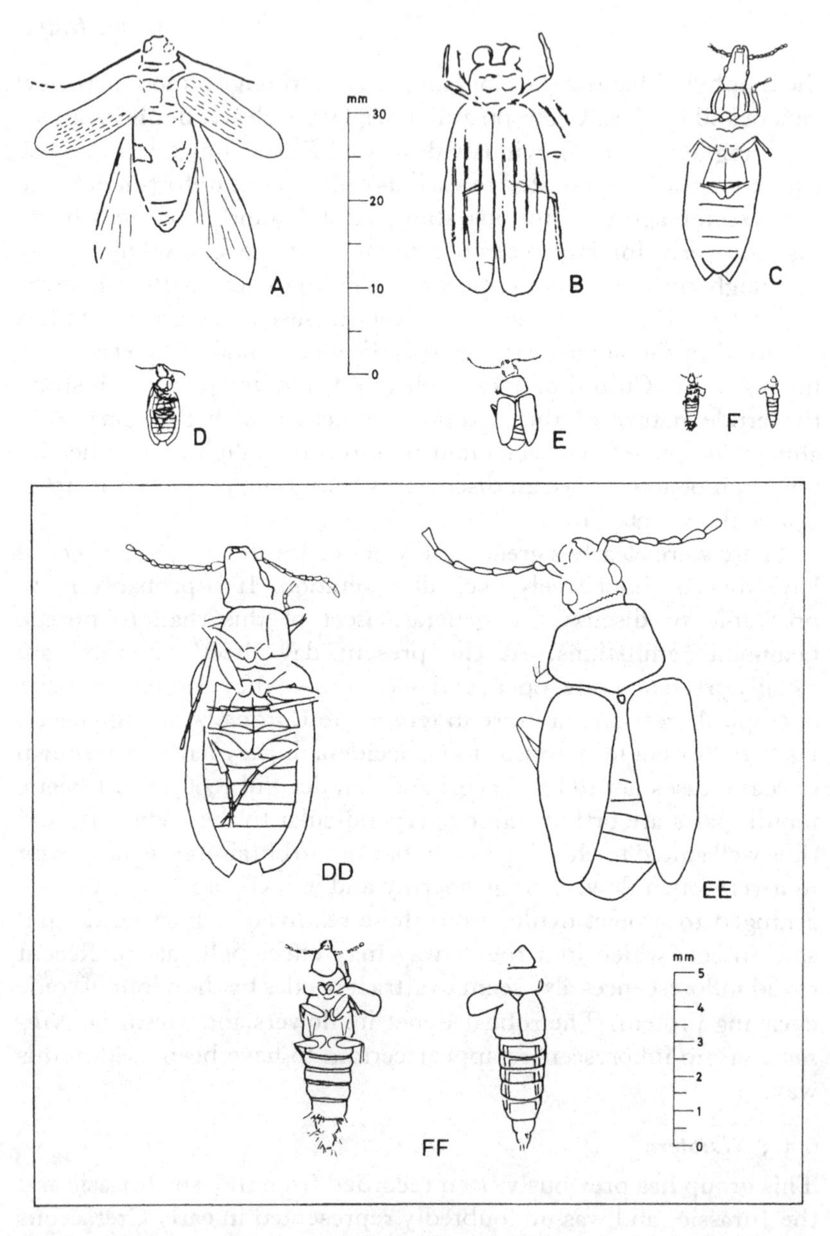

Figure 6.3. Diagrams of examples of late Jurassic Coleoptera, ×1 unless otherwise stated; A–D probably Cupedidae; E Chrysomelidae; F Staphylinidae. A, B from Lithographic Limestone, Bavaria; Kimmeridgian; after Handlirsch (1906–08). C–F from Karatau, Kazakhstan; late Jurassic; after Rohdendorf *et al.* (1968). A *Ditomoptera minor*; B *Opsis bavarica*; C *Tetraphalerus maximus*; D *Omma jurassica*; E *Protoscelis jurassica*; F *Archodromus brachypterus*. DD, EE, FF the same, ×4.

brate and other animals rather than plants. In early Cretaceous time Diptera (fig. 6.4) were probably either unimportant in pollenation or at most comparable with Coleoptera.

6.3.4 *Hymenoptera*

The numerous taxa recorded from Karatau and elsewhere are almost all (fig. 6.4) of the suborder Symphyta (sawflies and associated forms); at the present day in this group adults feed a little but the caterpillar larvae live by cutting into leaf edges in a characteristic manner which should be detectable in fossils. Of the Aculeata, one or two primitive wasps have been recorded in the late Jurassic and recently in the early Cretaceous (Evans 1969, Schlee and Dietrich, 1970). Ants (Smiley, 1969*a*) have now been recorded from late Cretaceous amber in north Alaska, but bees which are now such important pollenators in present-day temperate regions appear to be a group of Tertiary origin (Kelner–Pillault 1969).

6.4 OTHER ARTHROPODS AND MOLLUSCS

Spiders (Araneida) have not yet been described from the Mesozoic, although to judge from the numbers recorded from the Carboniferous and the Tertiary they must have been present as carnivores. Myriapods were also probably present.

Land pulmonate gastropods (Stylommatophora) are first recorded from late Cretaceous (Turonian), although aquatic pulmonates have a Jurassic or even a late Carboniferous record. Shell-less organisms of this group were presumably a later development.

6.5 ANIMAL INTEGRATION

The degree of integration of animal with plant life in early Cretaceous time was almost certainly low; it is probable that all the elaborate pollenation arrangements at the present day, concerning bats, bees, birds, butterflies and moths, have all developed with the angiosperm radiation in Tertiary time. Only Coleoptera (beetles) and perhaps some reptiles were of importance before that; the latter can potentially be investigated through coprolites (e.g. Harris 1945, 1964) but these fossil excreta are not often found.

It was presumably the already established angiosperms with large laminate short-lived leaves which provided the possibility for radiation of mammals with their very high feeding rates. Whether the

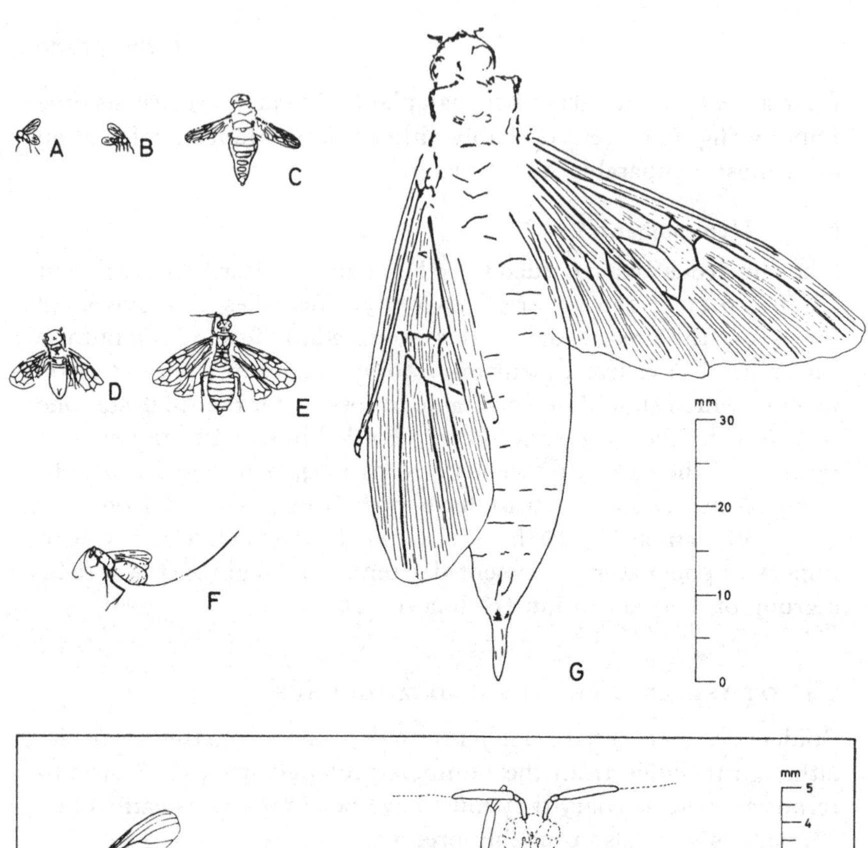

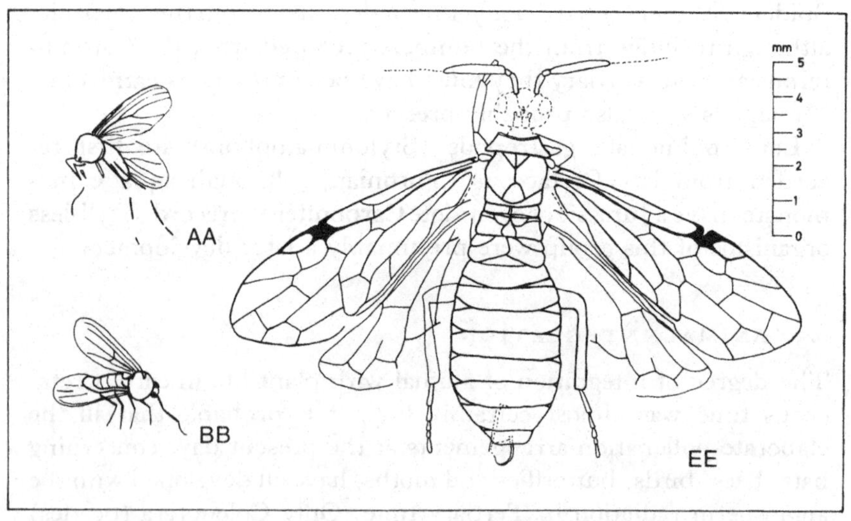

Figure 6.4. Diagrams of examples of late Jurassic Diptera (A–C) and Hymenoptera–Symphyta (D–G), ×1 unless otherwise stated. A, B from Purbeck Beds, Vale of Wardour, England; 'Tithonian'; after Brodie (1854). C–E from Karatau, Kazakhstan; late Jurassic; after Rohdendorf *et al.* (1968). F, G Montsech (Spain) and Solnhofen respectively; Kimmeridgian; after Handlirsch (1906–08). A *Thimna defossa*; B *Hasmona leo*; C *Protonemestrius martynovi*; D *Prolyda karatavica*; E '*Angaridyela vitimica*' (reconstruction); F *Ephialtites jurassica*; G *Pseudosirex schröteri*. AA, BB, EE the same, ×4.

late Cretaceous Ceratopsian dinosaurs benefited in this way is not known, but they were at least quite distinct from other earlier dinosaur herbivores, some of which may have fed entirely on aquatic vegetation. The thick-cuticled and small-leaved Jurassic and early Cretaceous gymnosperm trees may well not have been a major food source at all.

One of the strangest phenomena is the succession of attempts by palaeontologists and others (e.g. Urey 1973) to explain extinctions of dinosaurs through catastrophic cosmic events. Biologic explanations or direct continuing physical causes, such as unusually high temperatures at the end of the Cretaceous seem much more likely although less dramatic. In fact the high temperatures indicated for the Cretaceous seas by oxygen isotope studies on belemnites are perhaps the main guide to all the unusual evolutionary expansion of land plants and animals that took place together from Albian through Palaeocene time.

7 Cretaceous stratigraphy

7.1 PURPOSES OF STRATIGRAPHIC STUDY

Some exploration of Cretaceous stratigraphy can guide the search for new palaeobotanical evidence, particularly for assemblages of palynomorphs and fragment fossils. More immediate however is the problem of time-correlation of existing and future evidence; this depends on the existing ill-defined (traditional) time-scale, and on bio-correlation methods which are particularly undeveloped in the non-marine strata containing the majority of relevant fossils.

7.2 CRETACEOUS RADIOISOTOPIC DATES

Casey (1964) was able to estimate an age in years for the beginning (135 Ma ago) and the end (65 Ma ago) of the period. He could not give dates for the subdivision boundaries and suggested that the duration of each scale age should be estimated at 5–6 Ma, so that the Albian–Cenomanian boundary would be estimated as about 100 Ma. All such figures should be understood to fall within a bracket to cover the experimental uncertainty and also in almost all cases the geologic uncertainty as well. Unfortunately such figures have the attraction of simplicity and there is an understandable tendency to use them alone; they are at present even less meaningful than the as yet not properly defined scale divisions.

7.3 TRADITIONAL CRETACEOUS TIME-SCALE

In common with other geological periods the Cretaceous period has a traditional stratigraphic sequence scale based on the rock succession in western Europe (mainly in France). This scale for the period is divided into twelve stages (fig. 7.1) which were conceived originally as containing characteristic sets of fossils; the interchange of terms between the so-called 'time–rock unit' (stage) and an equivalent division of a time (age) may be regarded as merely

	Age/stage	Ma	District of current definition	Principal megafossil floras
	Palaeocene		Belgium	
		65±2		
Cretaceous Period	MAESTRICHTIAN		S.E. Netherlands	
	CAMPANIAN		N. France	
			* * * *	
	SANTONIAN		S.W. France	Magothy
	CONIACIAN		S.W. France	Kome?
			* * * *	
	TURONIAN		Central France	
				Raritan
	CENOMANIAN		W. France	Peruč, Blairmore?
		(100)		Potomac, Corwin?
	ALBIAN		Central France	Kolyma, Suchan?
			* * * *	
	APTIAN		S. France	Lower Greensand, Lena
	BARREMIAN		S. France	Franz Josef Land
	HAUTERIVIAN		W. Switzerland	Quedlinburg
	VALANGINIAN		W. Switzerland	
	BERRIASIAN		S. France	English Wealden
		135±5		
	'Tithonian'		S. France	
			* * * *	
	Kimmeridgian		S. England	

Figure 7.1. Table showing sequence of divisions (ages/stages) of the Cretaceous period, and the district of origin of each divisional name. * * = change of basin of deposition. The age of the main assemblages of the principal megafossil floras is shown.

semantic or as of philosophic importance according to taste, but the actual names such as the Albian age/stage remain the same. There was no original intention, nor indeed any method, of obtaining equal duration for these divisions and the incidence of successful radiometric datings of the actual strata, as opposed to discordant igneous rocks, has not been sufficient to alter the position materially in the Cretaceous period.

The names of ages/stages given in fig. 7.1 are of those recommended for use, and it seems advisable to eliminate unnecessary additional names or ranks, which often cause confusion, from the hierarchy. One such name is *Senonian* which refers to Coniacian+Santonian+Campanian+Maestrichtian and has now become superfluous. Another such name is *Neocomian* which has been widely used in America, Australia and elsewhere to mean simply 'early Cretaceous', but which appears to grant a false precision; its original meaning in western Switzerland was Valanginian+Hauterivian but its users often appear to mean Berriasian to Barremian inclusive, or even some unstated variant of this. More difficult to avoid are Volgian and Ryazanian which are approximately equivalent to Tithonian+Berriasian; the names are valid and useful as parts of a Regional Stratigraphic Scale (see Harland *et al.* 1972) for European Russia and neighbouring areas, but some stratigraphers (not particularly those from the USSR) prefer them to the French names for the World Standard Scale; although this difference of opinion has not yet been formally resolved the practical arguments for using these Russian names in more than a regional sense seem unlikely to prevail. Whatever the apparent difficulties of definition in the calcareous facies of the original Tithonian and Berriasian, agreement should be sought to erect age/stage and period/system boundaries in as low palaeolatitudes as possible so that the chance of correlation from the rocks of either palaeohemisphere is favourable; this would rule out Volgian and Ryazanian.

7.4 PREPARATION OF THE STANDARD STRATIGRAPHIC SCALE

The Stratigraphy Commission of the International Union of Geological Sciences (IUGS) has initiated the formal registration of point-marked boundaries, which will only be changeable by going through a similar full international procedure. Neither the beginning nor the end Cretaceous boundaries have yet been agreed but the former is under discussion; and a Cretaceous sub-commission, which has recently been formed, will be able to deal in the same way with the age/stage and other boundaries.

While the ages/stages were still identified by their fossil content, their use was equivalent to stating that something was known about the middle or core of each one. What was not known about them was the placing of their boundaries which had usually been taken for original convenience within the time ranges of the deposition-failures known as unconformities. As a result the boundaries are usually quoted as lying at the base of a zone named by an ammonite (fig. 7.2), although a first occurrence of one ammonite often has little time significance if it follows a gap in the succession of fossils. As

Age/stage	Definition taken at base of:	Difficulties
TURONIAN (Touraine)	Mammites nodosoides Zone Inoceramus labiatus Zone	Overlap of body-stratotypes; rarity of Chalk ammonites
	(?)	
CENOMANIAN (Le Mans)	Mantelliceras mantelli Zone	Unconformity at stratotypes base; condensed succession in England
	(dispar)	
ALBIAN (Aube, Paris Basin)	Leymeriella tardefurcata Zone	Change of province
	(nodosocostatum)	
APTIAN (Vaucluse–Provence)	Prodeshayesites fissicostatus Zone	South France sequence not elsewhere recognisable
	(recticostatus)	
BARREMIAN (Basses-Alpes–Provence)		

Figure 7.2. Definition of the beginnings of the stages Aptian to Turonian; last chronozone of each stage shown in brackets ( ). The third column shows the nature of the difficulties encountered at each boundary.

time-correlation methods improved it became necessary to define these boundaries accurately (see discussion in Harland *et al.*, 1967, explanatory preface) at least to avoid wasteful discussion about their placing. This work of definition referred to above consists principally of officially re-placing the necessary boundaries, first for periods and then for ages, in new regions away from the traditional name-bearing areas. The purpose of the change is solely to obtain an uninterrupted through-succession of rock deposition in which to place a single boundary point for reference. Because this work has not yet been completed all the scale boundaries in fig. 7.2 are for the present shown by broken lines to emphasise their lack of official agreed definition. It is consequently unlikely that correlations with the stratigraphic scale made at any great distance (or rather palaeodistance) from western Europe will have much precision until the definition of the scale itself is improved in the way suggested.

Sub-age	Chron/chronozone
UPPER ALBIAN	Stoliczkaia dispar
	Mortoniceras inflatum
MIDDLE ALBIAN	Euhoplites lautus
	Euhoplites loricatus
	Hoplites dentatus
LOWER ALBIAN	Douvilleiceras mammillatum
	Leymeriella tardefurcata

Figure 7.3. Divisions of the Albian age/stage in north-west Europe.

Because much of this definition work is in progress or in mind, the most important point to clarify is that no scale boundary should be related to any specific palaeobiologic event. The choice of a boundary-point should always be made well within a reference rock section of as uniform clastic lithology as possible with the maximum variety of available fossils and of other characters used or believed to be usable in correlation. This will provide a maximum number of correlation potentialities both above and below a boundary-point selected either purely for convenience or for supposed closeness to the traditional (but undefined) scale level.

7.5 TIME-CORRELATION METHODS

As exemplified by the Albian age (Fig. 7.3) the smallest formal divisions of the Stratigraphic Time-Scale are the chrons (or standard chronozones) which in this case for historical reasons still bear the names of prominent ammonites. Each of these ammonite names such as *Hoplites dentatus* was originally selected to represent the whole ammonite fauna of an assemblage biozone and most published correlations are based on this.

The major difficulty is that ammonites like most other fossils are in practice restricted to certain faunal realms or provinces. Biozone sequences will therefore differ and the sequences in each province require correlation through the margins of adjacent provinces. The quotation therefore of rocks in Australia or the Soviet Far East as, for example, Middle Albian, which can only be meant in the sense of the west European succession, represents a tenuous argument most difficult and costly to check. It is not suggested that such datings are grossly wrong but that before too much is read into diagrams such as are given in figs. 5.6 and 5.7, a difficult assessment must be made.

7.6 SEARCH FOR FURTHER PLANT MATERIAL

Palaeolatitude maps for geological periods are now being made from palaeomagnetic data with increasing accuracy (e.g. Smith *et al.* 1973). It is thought that with the help of ocean floor evidence the origins of the present tectonic plate patterns are interpretable back to Triassic time, so that for the Cretaceous there is reasonable confidence in these results. This means that for land plants the palaeolatitude of the flora now becomes something 'known' from which information it may be possible to begin to make palaeoclimatic interpretations from plants (see Hughes 1973*a*) in the Mesozoic; previously this was an increasingly difficult speculation as familiar types of plant were absent even in the early Tertiary and out of the question in the Mesozoic (Hughes 1963*a*).

On the maps of Smith *et al.* (1973) it will be seen that there are considerable changes in the position of North America and the eastern USSR from Jurassic to Cretaceous and again to Eocene time. The general effect of this in the USSR is to make the southern boundary of the Cretaceous Siberian floral province of Vakhrameev *et al.* (1970) parallel to lines to latitude and thus not calling for any other special explanation. In Western North America, the Cretaceous latitude of Calgary was 65° KrN, with British Columbia even higher and Alaska polar.

The Cretaceous equatorial region is poorly represented by available outcrops and fossils, as it happens to cross several countries now which for various reasons are among the less well explored stratigraphically.

7.7 EROSION AND SEDIMENTATION

The old ideas of cyclic occurrence of orogeny in time, which left the Jurassic and early Cretaceous as non-orogenic periods in the world, should have been corrected by the consequences of plate tectonic theory if not by observations on the geology of Pacific and Tethyan margins. There must at all times be plate destruction margins and some of them will inevitably involve orogeny, uplift and the resulting sedimentation; in early Cretaceous time this was occurring at least in parts of the north side of Tethys, in Western North America and in Grahamland area of the Antarctic.

Transgressions of the sea have been invoked as causes of various sedimentation and palaeobiologic phenomena, with particularly persistent emphasis on one in the Cenomanian. As far as is known there was no glaciation at this time to have provided eustatic sea-level movements, and no other cause for eustatic rises unless some entirely improbable 'sub-crustal bubble' is postulated. As far as the rock evidence is concerned, the idea of a Cenomanian 'transgression' appears to have originated in western Europe and North Africa perhaps with the French geologists; even in south-west England the relevant activity was Albian. In other places 'trangressions' occur in all other ages and no world-wide pattern is discernible; they are really better regarded as epeirogenic 'immersions' of parts of continental edges and therefore of regional significance or less.

An exceptional sedimentation was the European Chalk; it is only necessary here to record that its existence and duration should be fitted into any evolutionary pattern of tectonic plate history.

7.8 PALAEOTEMPERATURES

The well-known ^{18}O determinations of palaeotemperatures from belemnites give apparently high values for Santonian seas and much higher temperatures than now in high latitudes (Lowenstam 1964). The absolute values are uncertain, but they were strikingly high by comparison with the present-day glacial or inter-glacial conditions. It has unfortunately proved difficult to follow up this work in less suitable environments and with less suitable fossils than belemnites; this is illustrated by a recent discussion (Allen *et al.* 1973) of aragonitic shells from the Wealden of England which were said to yield ^{18}O-palaeotemperatures 'broadly consistent with warm temperature–subtropical conditions'; the relevant palaeolatitude of southern England would have been about 30° KrN.

Although it involves considerable extrapolation, the most useful simple hypothesis seems to be that of a continuous temperature rise from a Jurassic period already warmer than now to be maximum (Radmax) near the end of the Cretaceous (fig. 7.4) which may have been the hottest period at any time since land-plants appeared. Subsequently throughout the Tertiary, at least from Eocene time onwards, the temperatures were clearly falling towards the Quaternary glacial circumstances.

Unfortunately all the first work on palaeotemperatures referred to above was written up in terms of present-day latitudes before plate movement theory was developed; much of the argument presented is therefore misleading. Although there were considerable anomalies recorded in the work, particularly with observations from the USSR, the observed facts appear to fit better when Cretaceous palaeolatitudes are used. Such reassessment cannot perhaps be carried out with real profit until interpolated palaeolatitude maps (between 'mid-Cretaceous' and 'Eocene' of Smith *et al.* 1973) can be provided for each relevant age/stage in approximately the same stratigraphic detail as that in which the fossils used were described.

7.9 'RADMAX' HYPOTHESIS AND PALAEOBIOLOGY

One of the more remarkable characteristics of the earth has been the preservation for more than 3000 million years of surface temperatures in the narrow belt of tolerance of carbon-based life. Within that range, it is clear that on several occasions glaciation has denied large sectors of the crust to most organisms; the chief criterion for recognising these occasions is the physical nature of tillites and of other sediments. It is also clear that the temperature fluctuation to cause ice-spreading from high latitude mountain areas is small, through a critical range. It seems reasonable to suppose that opposite fluctuations could also occur, and that although no such obvious universal sedimentation phenomenon as tillite could be expected, an overall increase of calcareous deposition might be an indication; otherwise in this case the chief evidence might be biological. The hypothesis here advocated (fig. 7.4) is that numerous late Cretaceous events fit this pattern well and that the pattern may now well explain various suggested difficulties (see Hancock 1967).

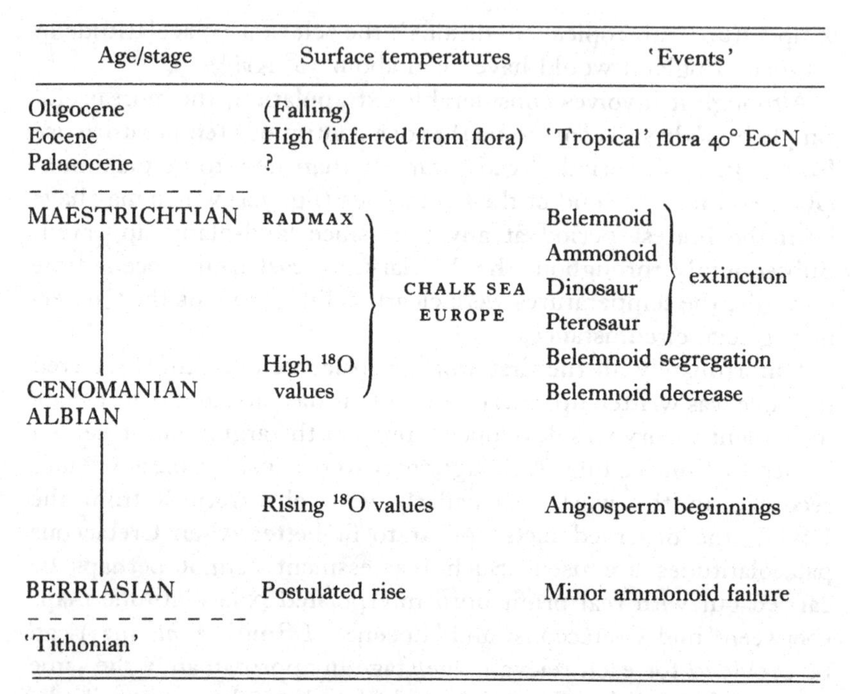

Age/stage
Surface temperatures
'Events'
Oligocene
(Falling)
Eocene
High (inferred from flora)
'Tropical' flora 40° EocN
Palaeocene
?
MAESTRICHTIAN
RADMAX
Belemnoid
Ammonoid
CHALK SEA EUROPE
Dinosaur
extinction
Pterosaur
High ¹⁸O values
Belemnoid segregation
CENOMANIAN
Belemnoid decrease
ALBIAN
Rising ¹⁸O values
Angiosperm beginnings
BERRIASIAN
Postulated rise
Minor ammonoid failure
'Tithonian'

Figure 7.4. Suggested surface temperature pattern for the Cretaceous period with postulated 'Radmax' in or just before Maestrichtian time.

(a) End-Cretaceous extinctions on land were very selective; dinosaurs and pterosaurs failed but birds and apparently mammals continued slowly diversifying; temperature regulation problems are indicated. Angiosperm plants diversified even more strongly, which appears to be a predictable response.
(b) End-Cretaceous extinctions in the sea were principally of the nektic cephalopods, e.g. ammonoids and most belemnoids; there may well have been some limiting factor concerning buoyancy. Extinction was gradual and the geographical range of belemnoids even became contracted into high latitudes towards the end of the Cretaceous; there were one or two survivors to a final extinction in the Eocene.
(c) Benthonic marine life was not much affected at the end of the Cretaceous, and some of the echinoid extinctions were perhaps facies controlled.
(d) Both organic shelled and calcareous plankton were involved in evolution rather than extinctions.
(e) Chalk Sea. The remarkable European Chalk deposition was

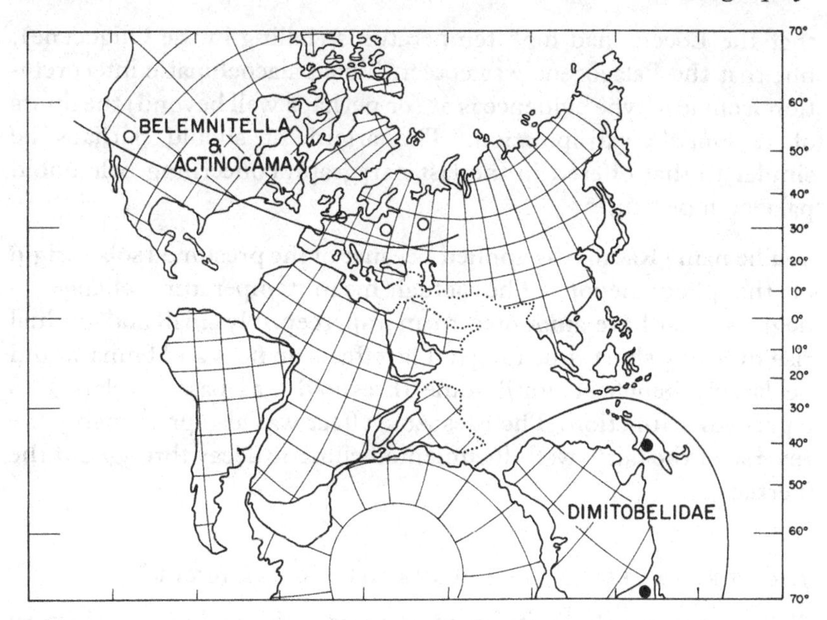

Figure 7.5. Distribution of post-Cenomanian Cretaceous belemnoids. North of 25° KrN, *Belemnitella* and *Actinocamax*; south of 25° KrS, Dimitobelidae.

probably both initiated and ended by incidental adjustments of plate movement concerned in the middle and north Atlantic ocean-floor spreading. It seems reasonable to believe that it was sustained by exceptional productivity of phytoplankton, but also the general history of coccolith and planktonic foraminiferan evolution supports a run-up of rising temperature from Jurassic time onwards.
(f) Both Jurassic and Cretaceous (and even early Tertiary) show less latitude contrast, particularly in land-plant assemblages, than either in Permian or in Recent time.

Although all kinds of further details appear to fit the general Radmax hypothesis of fig. 7.4, it is proper to record some which do not:

(g) The palaeotemperature work of Lowenstam (1964) and others he was reviewing indicated a drop of temperature (although not a very great one) in Campanian times. This may mean that there were complications comparable with interglacials, but also it would be best to await a re-assessment including that of the Cretaceous palaeolatitudes of the belemnite finds.
(h) Palaeobotanists have for long postulated (e.g. Barghoorn 1951)

that the Eocene had high temperatures (falling in the Oligocene), but that the Palaeocene was cooler. This palaeoclimatic interpretation from leaf-type evidence is at (or perhaps well beyond) the limits of reasonable extrapolation. The argument should perhaps be similar to that offered in the last paragraph concerning belemnoid palaeotemperatures.

The name Radmax is applied because of the presumed solar origin of the phenomenon. The actual mean temperature change in degrees must have amounted to an imperceptibly small and gradual rise over any short time range. The effects on flora and fauna would be largely beneficial until some threshold was passed relating to a particular function. The Mesozoic effect was also presumably the reverse of the fairly well-documented climatic decay throughout the Tertiary.

7.10 CRETACEOUS STRATIGRAPHIC 'CHARACTER'

Taking stratigraphy in its broadest sense as the evolutionary history of the earth, there is nothing exceptional about the Cretaceous period of that evolution. The invocation of a hiatus in orogeny, of world-wide transgressions or of great external catastrophe (Urey 1973) are entirely unnecessary to explain the known facts; such dramatic effects have surely become far less likely to be substantiated in view of the continuous evolutionary aspects of plate tectonic theory. They have usually been invoked by both palaeontologists and non-palaeontologists to account for apparent disharmonies such as the origin of angiosperms, the extinction of dinosaurs and other organisms, or the existence of chalk deposition. However, consideration of these many events has led to the conclusion that all of them can be accepted as part of a normal evolutionary pattern subject to one progressive late Mesozoic external factor; this factor is the gradual rise of mean crustal surface temperature of the earth through Jurassic and Cretaceous time to a maximum near the end of the Cretaceous, referred to in the last two sections. The actual mean temperature change in degrees must have been a very small gradual rise; in the same way the change for the reverse process of Quaternary glacial advance is believed to have been minimal through the critical range, and also to have been a part of a radiation change which ran throughout the Tertiary and doubtless continues.

PART 3

Critical fossil evidence

8 Jurassic gymnosperms

All mid-Mesozoic plants are separated from their present surviving successors by the generations involved in 150 million years. As might be expected from the geological evidence of relative similarity of certain fossil and Recent habitats, some individual plant organs resemble in certain characters those of living plants, but there is no proof at all of whole plant resemblances. As has been pointed out by Harris (1961, p. 322) and others who have built up great knowledge of Mesozoic plant fossils, the chief feature of such study is that so much detail is as yet unknown and nearly all of the synthesis unattempted.

8.1 MODE OF PRESENTATION OF PLANT GROUPS

Jurassic seed-plants are here presented as far as possible without reference to the classification of living plants. They are therefore discussed in an order based on general leaf form, this being the most prominent feature of the most frequently occurring megafossil organ; the generic or other name beginning each paragraph may be used as a group name if required. This approach is not used in order to deny affinities already suggested but rather to leave an open field for a Mesozoic classification based on all available characters weighted equally, as a prelude to establishment of a true phylogeny which can then be of some stratigraphic importance.

Such fine distinctions as have already been made, e.g. the recognition of taxads by Florin (1958) and by earlier workers, were due to their good knowledge of the characters of living plants. While this is probably the only practicable way to arrive at distinctions and interpretations, it need not and should not affect the classifications of fossils. Harris (1953*a*) particularly has expressed pessimism about stratigraphic distinctions within the well-preserved Yorkshire Deltaic flora of the mid-Jurassic, but the time-span involved in that case was only a part of the Bajocian age/stage and was thus too short for such distinctions in view of the use at that time of a classification based largely on extant taxa.

The fossils mentioned by name below are all from the large well-known flora of the Deltaic Beds of East Yorkshire (Bajocian), unless specifically otherwise indicated. Triassic, and post-Bajocian Jurassic plants are only included when their groups are presumed to have existed but are not represented in the Yorkshire Bajocian flora. Selected characters and the extent of the fossil record are given for each group, with reference where appropriate to fuller descriptions in textbooks and in other sources.

8.2 SIMPLE-LEAVED PLANTS

These plants are usually referred to as ginkgophytes and coniferophytes but it is desirable for the purposes of this book to avoid these groupings and thus to avoid prejudice over possible courses of evolution.

8.2.1 *Leaves with dichotomous venation*

Baiera (fig. 8.1*A*) and *Ginkgoites* (fig. 8.1*B*, *C*) show different degrees of lobation of the leaf. Some Jurassic leaves show minimum lobation and in this character resemble leaves of a mature living *Ginkgo* and have even been attributed to the extant species; the cuticle characters of the Jurassic leaves and the living species may also be similar. With *Baiera muensteriana* there is an associated catkin-like male microsporangiate organ, and another (fig. 8.1*D*, *E*) with '*Ginkgo*' *huttoni* from the Upper Deltaics (van Konijnenburg-van Cittert 1971); but other reproductive structures are not yet known in the group except for some dispersed seeds of uncertain origin. Some dispersed pollen grains *Monosulcites minimus* known from Jurassic onwards are of the same general elliptical shape as those of Recent *Ginkgo*. On the other hand the long and short shoots of *Ginkgo* have not been demonstrated in the Jurassic, and the first fossil wood which is certainly Ginkgoalen is from the Eocene. These leaves are common Mesozoic fossils but it is entirely premature to reconstruct whole plants from their evidence, especially when there is only a single extant relict species for guidance.

Czekanowskia (fig. 8.2*A*, *B*) has leaf lobation which is so deep that the units become linear segments. The general character of the leaf cuticle resembles that of *Baiera*. The leaves are borne on short shoots but they have been shown (Harris 1951*a*, *b*) in three localities to be

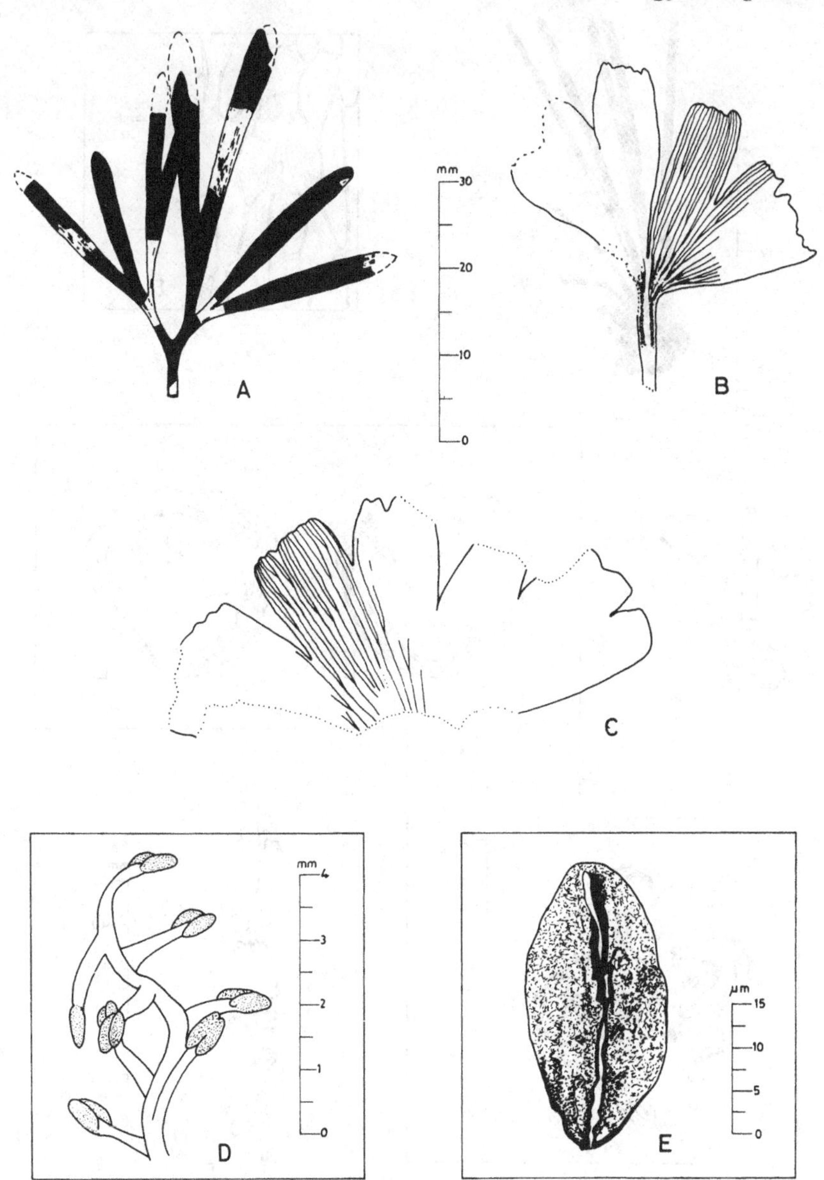

Figure 8.1. Jurassic 'Ginkgophytes', simple-leaved plants with dichotomous venation; ×1, unless otherwise indicated; Yorkshire, Bajocian. A *Baiera gracilis* Bunbury; leaf, drawn from Sedgwick Museum specimen K838, also figured Black 1929, fig. 8. B–E *Ginkgoites* leaves and associated specimens. B '*Ginkgo*' *huttoni* (Sternberg) Heer; leaf, after Harris 1948. C '*Ginkgo*' *digitata* (Brongniart) Heer; leaf, after Harris 1948. D male fructification associated with *G. huttoni*; ×7.5. E pollen from same, ×1000; both after van Konijnenburg-van Cittert 1971.

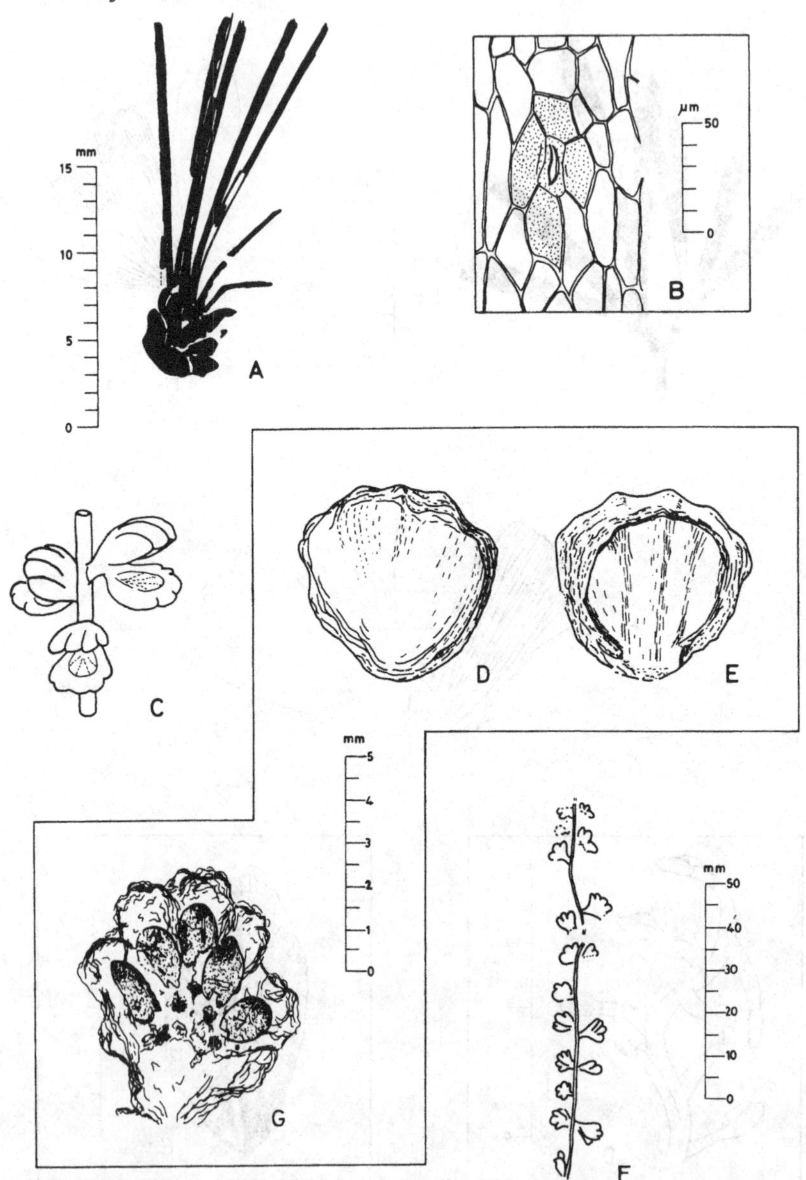

Figure 8.2. Jurassic Czekanowskiales; Yorkshire, Bajocian. A, B '*Solenites*' *vimineus*; shoot, ×2; after Harris (1951*b*); cuticle, ×250; after Harris (1951*a*). C–G *Leptostrobus* fructifications; after Harris (1951*a*). C–E *L. cancer*; Yorkshire, Bajocian. C reconstruction of part of cone, ×2. D outside of capsule valve, type specimen, ×5. E inside of same valve in balsam transfer. F, G *L. longus*, Scoresby Sound, East Greenland, early Jurassic. F part of cone, ×0.5. G capsule with five megaspores under each of which is a small mass of pollen, ×5.

closely associated with the *Leptostrobus* seed-bearing structure (fig. 8.2*C–G*). This discovery has removed *Czekanowskia* from the 'Ginkgoales' based on *Ginkgo*, but as Harris hinted it may at the same time remove all these Mesozoic leaves with dichotomous venation from such a group.

8.2.2 *Linear leaves with single vein or parallel venation* (*Linearphylls*)

The leaves mentioned are not expected to form a homogeneous group; they are best separated on cuticle details (see Florin 1958), but it is not yet by any means clear which cuticle characters are linked, which may show discontinuous variation, and what are the functions associated with these characters.

Elatides williamsonii (fig. 8.3*B*) from the Yorkshire Bajocian is one of the most fully described (Harris 1943) and discussed species. The leaves have a radiate arrangement on the branch, are square on cross-section and are amphistomatic, the cuticle is thin and the stomata lie in two well-marked bands running the full length of the under-side of the leaf; the stomata are arranged irregularly in the bands. A male cone with a matching cuticle is known from the same beds, and it has yielded *Perinopollenites elatoides* (fig. 8.3*C–E*) to Couper (1958) who regarded this as support for Harris' assignment of these fossils to the family Taxodiaceae; the female cone, however, does not support this in most details (fig. 8.3*A*). In fact the stomatal details are not exactly matched in that family either and, in view of the gaps in our knowledge of the complete plant, *Elatides* appears to be a good candidate for inclusion in a new Jurassic family that can remain uncommitted in affinity until its evolutionary history has been established from successive fossils. *Haiburnia setosa* Harris (1952) is another fossil (fig. 8.4*A*, *B*) from the Yorkshire Bajocian which has been similarly assigned and to which the same remarks apply. In both cases it is necessary to name the species under discussion because any apparent homogeneity in the genus may break down on removal of the implied affinity from consideration.

Elatocladus is a genus of small leaves to which Florin (1958) has added nine species (Yorkshire, Bajocian) based only on cuticles of leaf fragments. *E. cephalotaxoides* (Hettangian, Scania and East Greenland) has longer leaves and has been attributed to the Recent Cephalotaxaceae although its cuticle is unknown.

Thomasiocladus zamioides (fig. 8.4*E*), known on leaf and cuticle characters only, was attributed firmly by Florin (1958) to the family

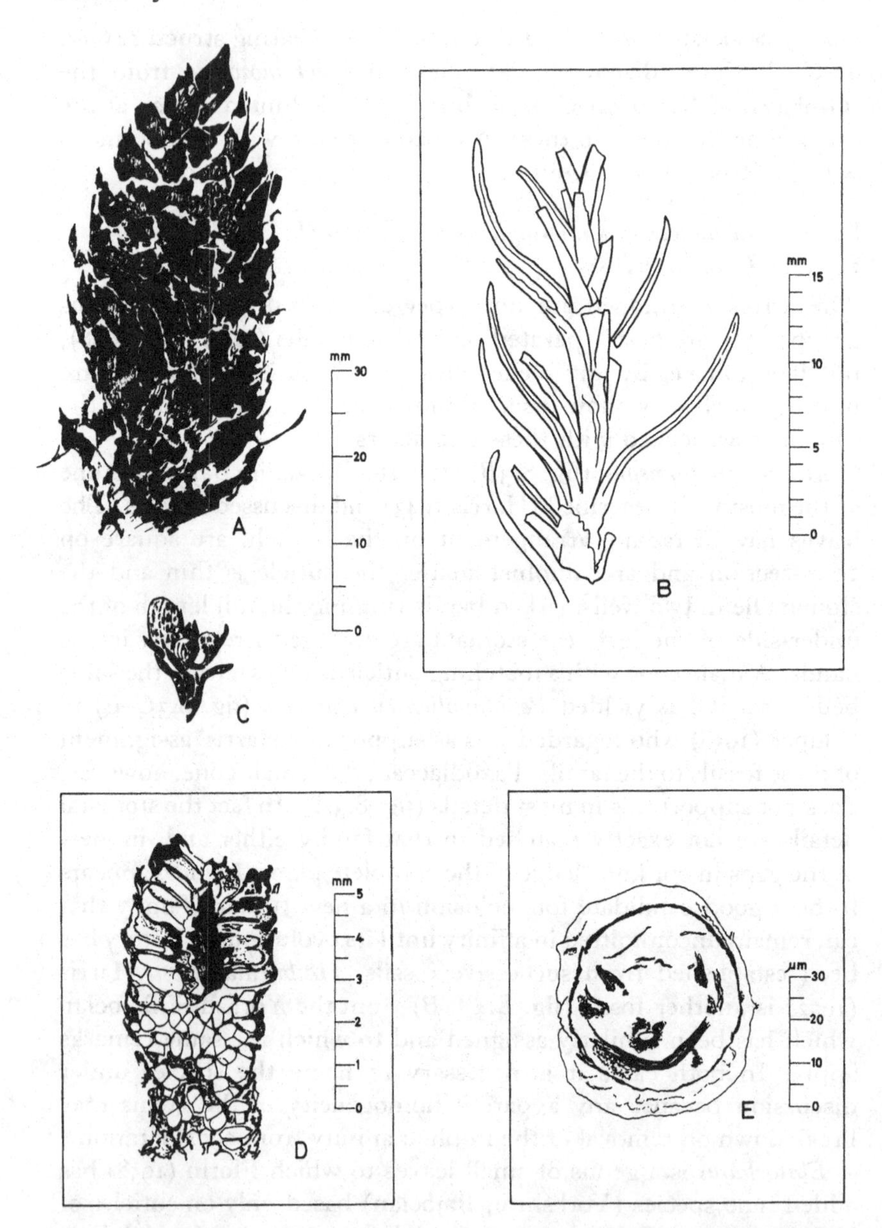

Figure 8.3. Jurassic Linearphylls; *Elatides williamsonii* (Brongniart) Seward, Yorkshire Bajocian; after Harris (1943). A female cone with some scale apices broken, ×1. B leafy shoot, ×2. C shoot with very young male cone, ×1. D oblique longitudinal section through unripe male cone. E pollen from cone, after Couper (1958), ×500.

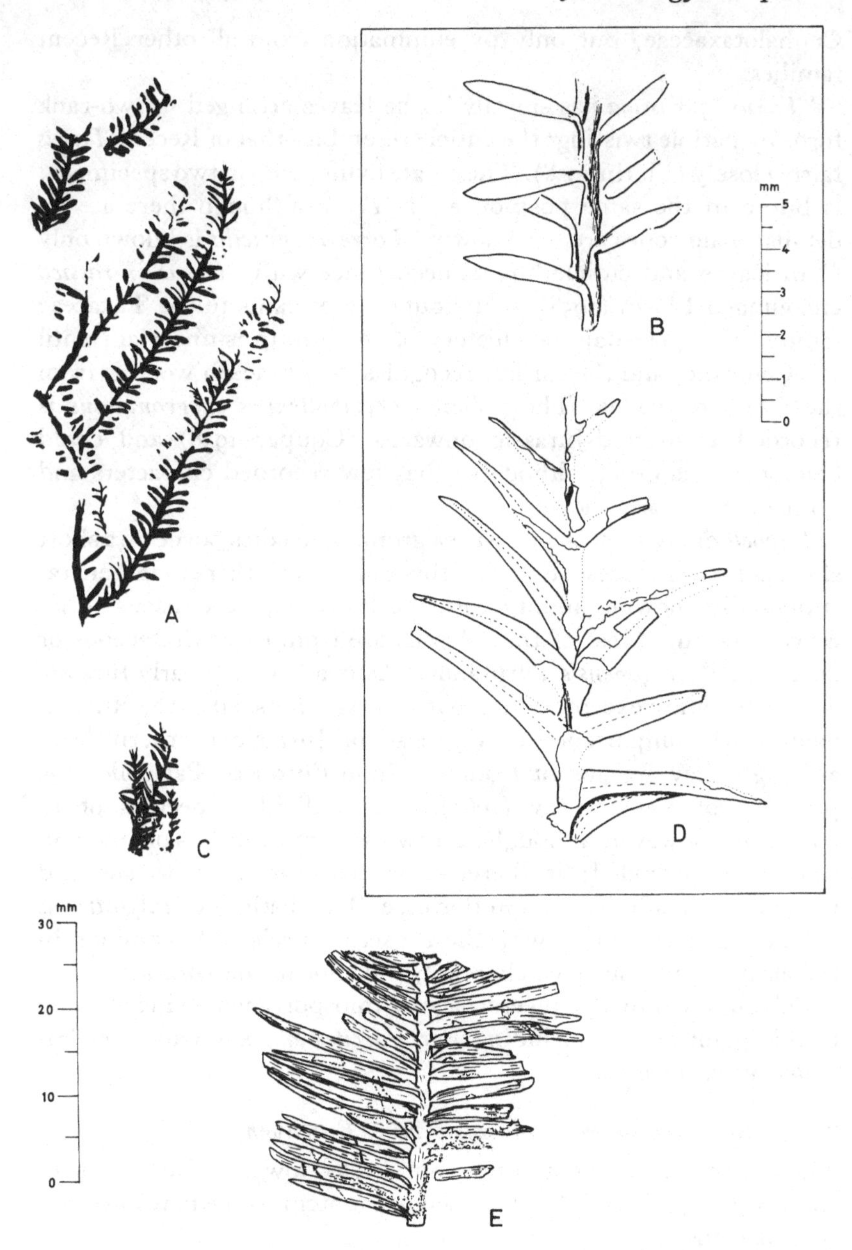

Figure 8.4. Jurassic Linearphylls (contd); Yorkshire Bajocian. A, B *Haiburnia setosa* (Phillips) Harris; after Harris (1952). A shoot, ×1. B detail of same, ×5. C, D *Elatides divaricatus* (Bunbury) Harris; after Harris (1951*b*). C shoot, ×1. D detail of same, ×5. E *Thomasiocladus zamioides* (Leckenby) Florin; branch with leaves, ×1; after Florin (1958).

Cephalotaxaceae, but only by elimination from all other Recent families.

'*Taxus*' *jurassica* has spirally borne leaves arranged in two-rank form by petiole twisting; the cuticle resembles that of Recent *Taxus* fairly closely (Florin 1958). The single ovule, seen in two specimens, is borne in the same position as in *Taxus* although there are no details; male cones are not known. '*Torreya*' *gracilis* is known only from leaves and cuticle but its occurrence with '*Taxus*' *jurassica* encouraged Florin (1958) to attribute these plants to the Taxaceae family. The post-Bajocian history of this group is uncertain until the Cenozoic, and the earliest recognisable Taxacean wood is from the late Cretaceous. The pollen *Spheripollenites subgranulosus* is recorded from mid-Jurassic onwards (Couper 1958) and could belong to this group although it has few recorded characters and no male cones are known.

Pityocladus is a leaf-genus with a group of needles borne on a short shoot as in the Pinaceae. Fossils of this and several other organ-genera, supposedly showing affinity with the Pinaceae, were discussed by Seward (1919). Most of these fossils are from early Cretaceous or later, but *P. ferganensis* Turutanova-Ketova from the early Jurassic is an interesting exception that has among others led to the Russian view of the importance of Pinaceae in Jurassic northern Asia, although they are absent from the Indo-European Palaeofloristic province of Vakhrameev (1964). The individual needles of *P. ferganensis* however, although narrow, were more or less planar, and were cross-wrinkled; it therefore seems reasonable to suspend judgement on affinity even in this case. The distinctive *Pityostrobus* of more certain affinity with the Pinaceae is relatively common in the early Cretaceous (see chapter 9) but not in the Jurassic.

Palynology provides several saccate miospores which may belong to this plant group, including *Parvisaccites enigmatus* and *Cerebripollenites mesozoicus*.

8.2.3 *Scale-like leaves with single vein* (*Brachyphylls*)

These leaves have smooth upper and keeled lower surfaces, and are spirally arranged, closely adpressed to the stem; some may have had several veins.

Brachyphyllum (fig. 8.5*A*, *C*) has leaves in which the free part is short, not exceeding the width of the leaf-base cushion. The leaf is amphistomatic with irregularly oriented stomata in many rows each one stoma wide, in a thick cuticle. *B. mammillare*, one of the commonest fossils in the Yorkshire Bajocian flora, was attributed to the Recent family Araucariaceae by Kendall (1949*c*); she demonstrated its association with the female cone-scale *Araucarites phillipsi* (fig. 8.5*E–H*) which resembles cone-scales of living *Araucaria* Section *Eutacta*; an interesting specimen of part of the cone was also recorded by Kendall (1952). The male cones (figs. 8.5*B, D*) are actually attached to *Brachyphyllum mammillare*, and the pollen was classified as *Araucariacites australis* by Couper (1958).

Pagiophyllum (fig. 8.6*D, E*) is separated only arbitrarily from *Brachyphyllum* by having the free part of the leaf longer than the width of the leaf-base cushion; it has a similar cuticle in many respects. Kendall (1952) attributed *Araucarites estonensis* to *P. connivens* on cuticle characters and on association, thus apparently extending the Jurassic Araucariaceae. The pollen, however, of the male cones of *P. connivens*, and incidentally also of *B. scotti* Kendall (1949*a*), is of *Classopollis* type (see Couper 1958) which has no Recent parallel (fig. 8.6*F, G*). *Classopollis* pollen is most directly associated with *Hirmeriella* from the Lower Lias (Jurassic, Hettangian) of western Europe (Hirmer and Hörhammer 1934; Harris 1957; Jung 1968), which has leaves like *Brachyphyllum* but which has been separated on the basis of its unusual female cone-scale. *Hirmeriella* appears to be a close descendant of the Triassic family Voltziaceae, although various authors wish to associate it with the Recent Podocarpaceae; it cannot be placed with the Araucariaceae from which it differs even in the traditional 'araucarian' character of crowded wood pitting. *Masculostrobus* from the Middle Jurassic of Iran (Barnard 1968) has also been shown to bear *Classopollis*.

The content of the last two paragraphs alone seem to constitute adequate argument for removing ideas of attribution to Recent families entirely from nomenclature and taxonomy of Jurassic fossils. Kendall herself (1949*b*), in describing *Brachyphyllum expansum* as 'completely separate from the Araucariaceae', pointed to this conclusion, as did Townrow (1967*c*), more positively, in separating his new Australian Jurassic genus *Allocladus* from *Brachyphyllum*; *Allocladus* leaves have a scalloped margin and are epistomatic, which is not true of any species of the other fossils mentioned in this section. Townrow (1967*b*), in describing another leaf of this kind, *Nothodacrium* from the Jurassic of east Antarctica, showed association with a '*Masculostrobus*' (*Pityanthus* according to Barnard (1968) who retains *M.* for those bearing non-saccate pollen) yielding pollen resembling *Tsugaepollenites*

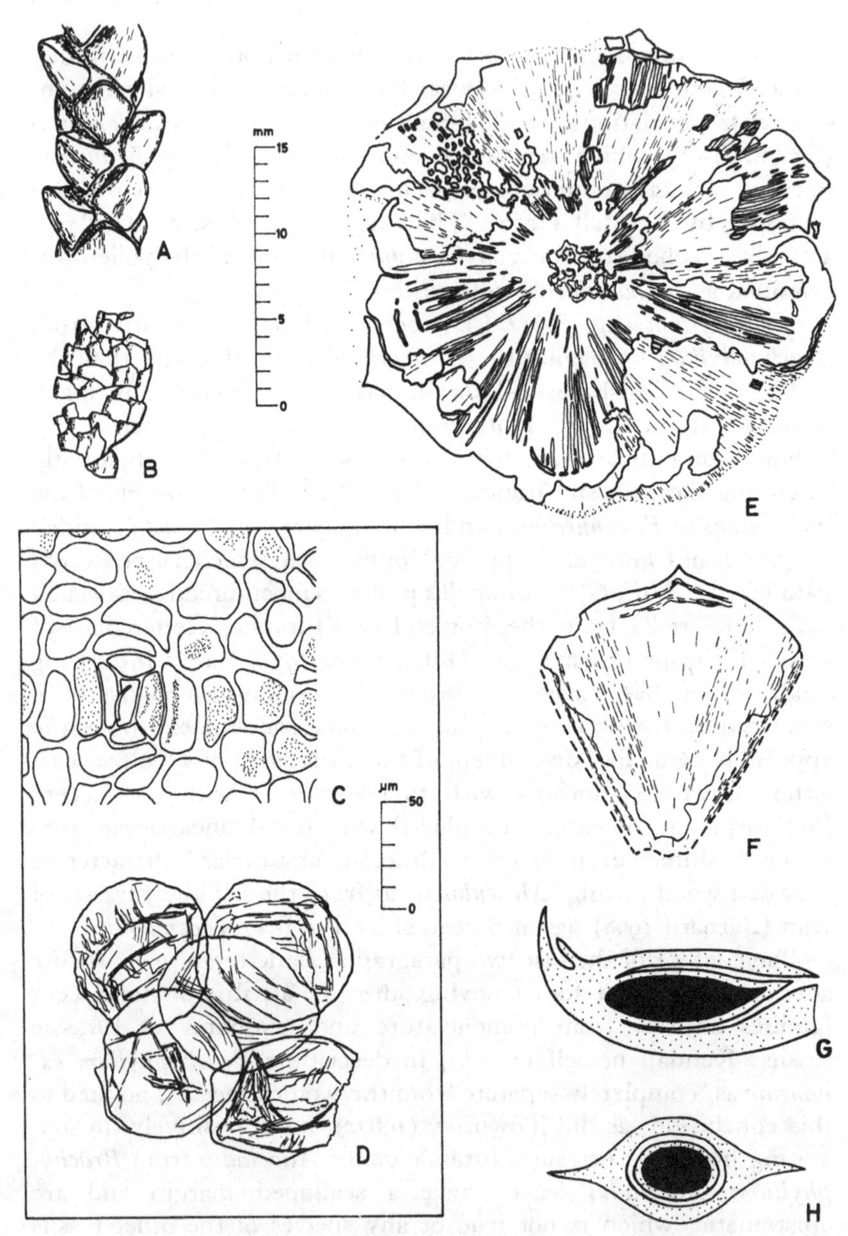

Figure 8.5. Jurassic Brachyphylls; Yorkshire Bajocian. A–D *Brachyphyllum mamillare* Brongniart. A restoration of twig, ×2; after Kendall (1949*c*). B male cone, ×2; after Kendall (1949*c*). C Leaf cuticle, ×250; after Kendall (1949*c*). D pollen from male cones, ×250; after Couper (1958). E–H *Araucarites phillipsi* Carruthers. E transverse section of female cone, ×2; after Kendall (1952). F–H cone-scale and reconstruction sections, ×2; after Kendall (1949*c*).

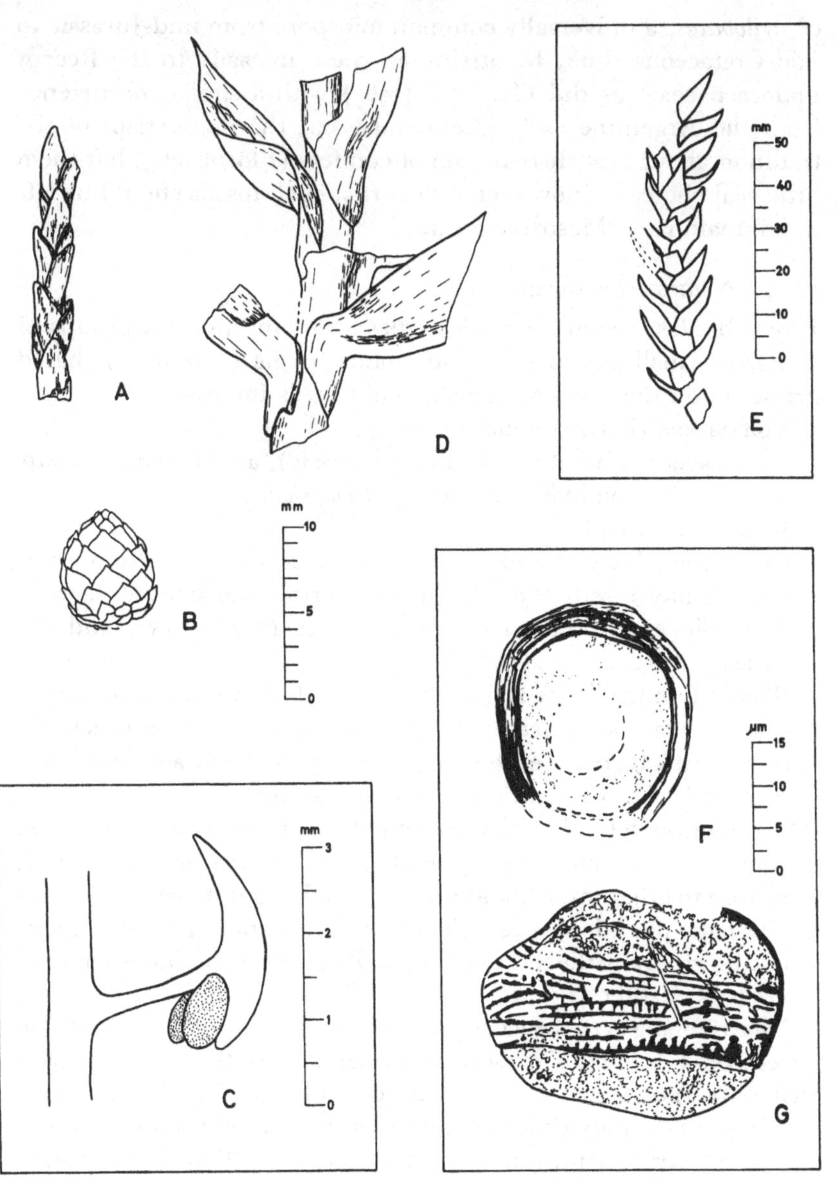

Figure 8.6. Jurassic Brachyphylls (contd); Yorkshire Bajocian. A–C, G, *Brachyphyllum crucis* Kendall. A shoot, ×2; after Kendall (1952). B, C, G after van Konijnenburg-van Cittert 1971. B possible reconstruction of male cone, ×2. C reconstruction of microsporophyll with pollen sacs, ×50. G Pollen grain, ×1000. D, E *Pagiophyllum insigne* Kendall; after Kendall (1948). D twig, ×2. E type specimen, ×0.5. F *Pagiophyllum connivens* Kendall (pollen grain), ×1000; after Couper (1958).

cf. *trilobatus*, a universally common miospore from mid-Jurassic to mid-Cretaceous time; he attributed these uneasily to the Recent Podocarpaceae, as did Gamerro (1965) with a similar occurrence from the Argentine early Cretaceous, but the dogmatism of the tradition about past distribution of conifers (Florin 1963) left them little real choice. It now seems clear that such fossils should be left in an unaffiliated Mesozoic group.

8.2.4 *Neutral classification*

From this last section it is clear that a classification for groups of Mesozoic small-leaved gymnospermous fossils ('conifers'), based primarily on the fossil sequence, could be as follows:

- Voltziaceae (Permian and Triassic)
- Palissyaceae (Late Triassic–Early Jurassic), as yet a small group
- Jurassic Brachyphylls (including *Hirmeriella*)
- Jurassic Linearphylls
- Cretaceous (new) families, based on fructifications, to accommodate any fossils thought to be intermediate between agreed Cenozoic extensions of Recent families (see below), and the Jurassic fossils.
- Recent families to be rediagnosed to include Cenozoic and even some Cretaceous fossils when the family has been satisfactorily traced back through each stratigraphic period and age concerned (but with extreme caution before Cretaceous Albian).

The intention behind this arrangement is to treat gross geologic age as the first character and generalised leaf form as the second; affiliation to other taxa should not be considered until several organs of the same plant have been identified. The names indicated above would be groups in the sense of a purely Palaeontologic Data-handling Code (see chapter 4).

Palaeobotanists of the early part of this century, and their German successors (Gothan, Kräusel), appreciated from the Jurassic fossils that were known that none of them could be fitted into any Recent families. The family Protopinaceae was erected (with a revived older name; see Kräusel 1919) for a purpose similar to that of the present suggestions above, in the separate but parallel field of Jurassic fossil wood; the name however was not adequately defined and it is better to allow it to lapse on these grounds of inadequacy because if used it could well be unduly misleading.

The consensus of palaeobotanical opinion a decade ago (see Florin 1958, 1963) was that Taxaceae, Taxodiaceae, Cephalotaxaceae and possibly Araucariaceae were the only Recent families that could be identified in any sense in the Jurassic. These are of course the smaller families, each of less than 40 species extant, which are now more or less composed of relicts. They contrast with the larger families Pinaceae, Podocarpaceae and Cupressaceae, each of 150 or more species, which are now agreed to have no pre-Cretaceous history. The crucial family for study in this connection is probably the Podocarpaceae, which has a pseudo-uniformity due to a shortage of generic names, and for which I join Townrow (1967*a*) in postulating a northern hemisphere or at least general equatorial origin; the timing of the 'migration' (or extension) to the south would, however, be in the Cretaceous, and not in the Triassic. Florin (1963) regarded the extensive East Asian occurrences of living Podocarpaceae as being a result of 'northward migration' in the Pliocene, but it seems just as likely that they represent undisturbed relicts of northern hemisphere or universal plants; several relict taxa of other plant groups in that general area appear to result from the general Cenozoic climatic decay down to the Pleistocene that affected the whole Eurasian flora. It should also be noted that the Cretaceous palaeolatitude of these areas was south.

8.3 PINNATE-LEAVED PLANTS

All these distinctive leaves were originally thought of as belonging to cycads; for sixty years however the majority have been known to belong to the Benettitales, the first plants with flowers.

8.3.1 *Pinnate leaves with haplocheilic stomatal subsidiary cells*

This group is inhomogeneous but as the reproductive organs are only known in the cases indicated (see Harris 1964), no further subdivision is attempted.

Nilssonia species include those like *N. tenuinervis*, with an unsegmented lamina, which was classified in the old form-genus *Taeniopteris*; the order Taeniopteridales is now restricted to leaves of this type from Triassic Karnian and earlier (Alvin *et al.* 1967). The well-known *Nilssonia compta* (fig. 8.7*B*) from the Gristhorpe Bed has a segmented lamina although the segments vary in width; it has been attributed by cuticle comparison to the same plant (fig. 8.7*A–F*) as *Beania gracilis* (female cone) and *Androstrobus manis* (male cone), and Harris (1964, p. 166) discussed such doubts as remain about this important hypothesis. *Ctenis* species have long pinnae with several

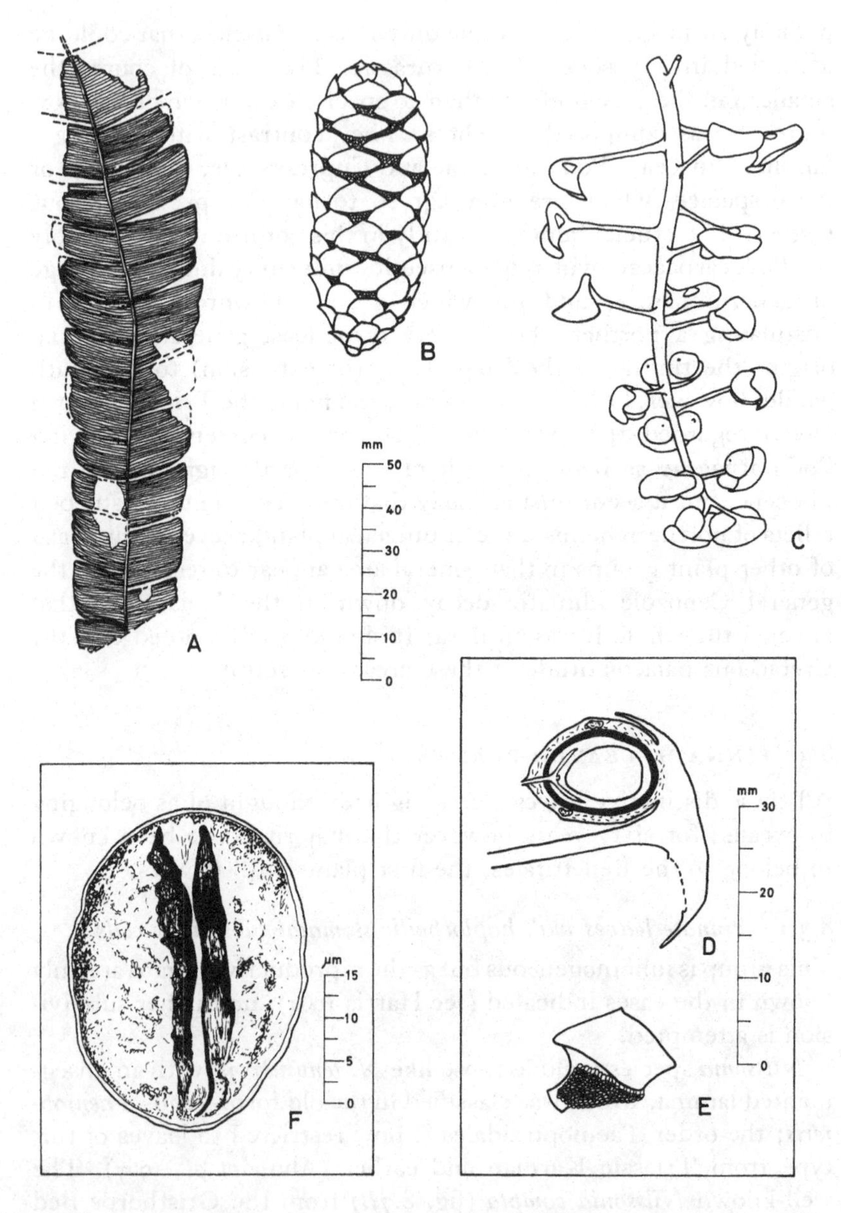

Figure 8.7. Jurassic Nilssoniales (cyadophytes); Yorkshire, Bajocian. A *Nilssonia compta* Phillips; leaf drawn from specimen, ×0.5. B, E *Androstrobus manis* Harris; restoration of cone, ×0.5; restoration of single sporophyll, ×1; after Harris (1964). C, D *Beania gracilis* Carruthers; restoration of pendulous cone, ×0.5; reconstruction of megasporophyll in section, ×1; after Harris (1964). F *A. manis* (pollen grain), ×1000; after Couper (1958).

parallel veins which anastomose occasionally, but no reproductive organs are known. *Pseudoctenis* species differ in lacking anastomoses of the veins; *Androstrobus prisma* is attributed to the same plant as *Pseudoctenis lanei*. The pollen taken from *Androstrobus manis* is oval monosulcate of 35 μm diameter or less (fig. 8.7*F*); the pollen from *A. prisma*, however, is circular in amb, apparently inaperturate and finely granular (see van Konijnenburg-van Cittert 1971). A newly described leaf-genus *Paracycas* Harris 1964 is remarkably like Recent *Cycas*, and all these fossils have been taken to form part of a large Jurassic group of true cycads of which the nine surviving genera are scattered relicts. Less certain is the growth habit of the fossils, as none of the stems is known (Harris 1961). All of the fossil plants mentioned in this paragraph have relatively thin cuticles.

Pachypteris, as suggested by the name, has thick cuticles. *P. lanceolata* is bipinnate, and *P. papillosa* simply pinnate (fig. 8.8*B–D*), but the distinction is also clear on the details of the cuticle. *Pteroma thomasi* (fig. 8.8*A*, *E*) is a newly described male organ, attributed only by association, but with great care (Harris 1964, pp. 175–8), to *P. papillosa*; the bisaccate pollen (fig. 8.8*F*) resembles *Alisporites* (*Pteruchipollenites*) *thomasi* (Couper 1958) in that the sacci are lateral rather than distal, but the infra-reticulum of the sacci is very fine and thus distinct. Similar pollen has been taken from South African *Pteruchus* (see Townrow 1962) of the Corystospermaceae, which family has been grouped loosely with others under the heading 'Mesozoic pteridosperms'; *Pachyteris* is thus by implication similarly attributed, and it is now most important to find the female reproductive structures; it has been noted that *Pachypteris papillosa* only occurs in strata containing dinoflagellates, from which these strata are presumed to be marine (van Konijnenburg-van Cittert 1971). Also in the Yorkshire Bajocian flora are species of *Stenopteris* which have thick cuticles but are distinguished from *Pachypteris* in their very narrow ultimate segments that have only one vein; no reproductive structures are known. All these leaves are relatively similar to *Thinnfeldia* of the European early Jurassic and to *Dicroidium* of the late Triassic of the southern hemisphere, the latter being the main leaf-genus of the Corystospermaceae.

8.3.2 *Pinnate leaves with syndetocheile stomatal subsidiary cells*

These leaves are of the same general type as those mentioned in the last section, even to the inclusion of 'Taeniopterid' leaves with an unsegmented lamina. However, once the cuticle difference was

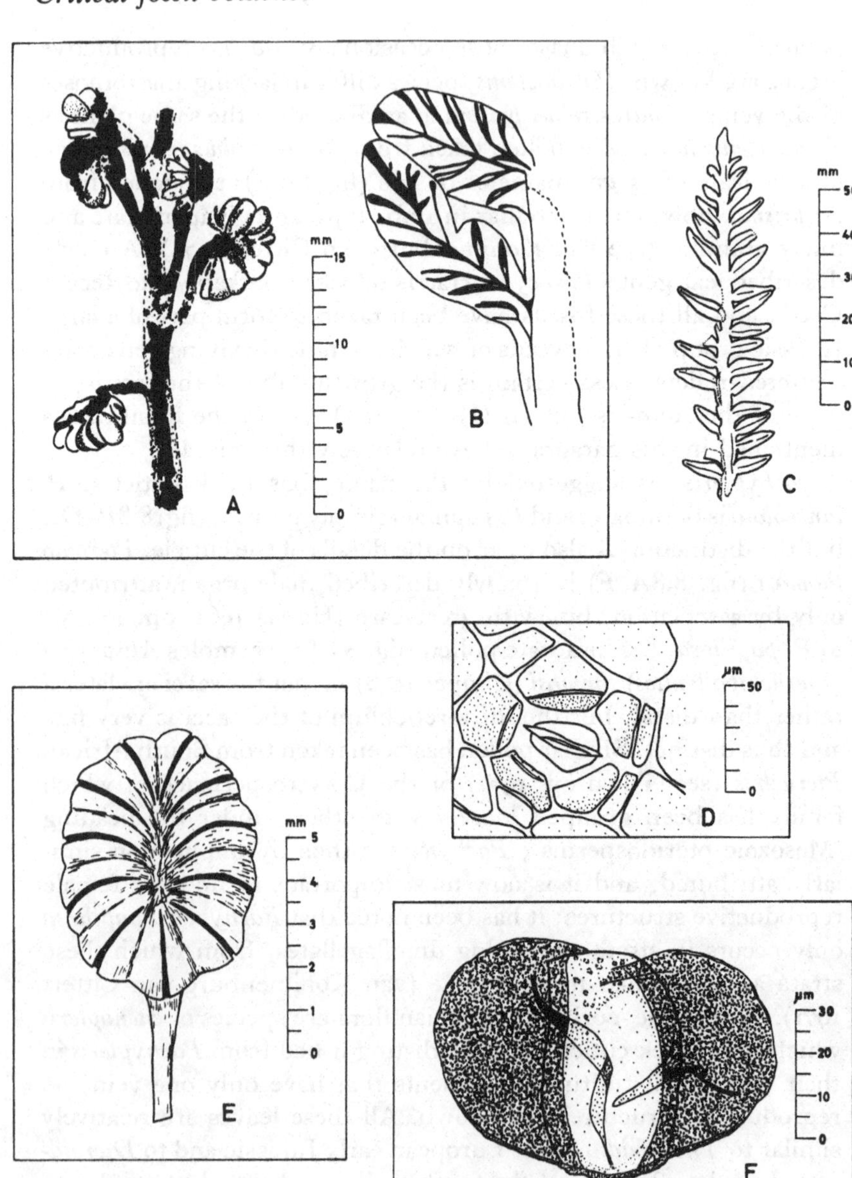

Figure 8.8. Jurassic 'pteridosperms'; Yorkshire, Bajocian; after Harris (1964). A, E *Pteroma thomasi* Harris. A middle region of holotype, ×2. E restoration of fertile head, ×5. B–D *Pachypteris papillosa* (Thomas and Bose). B leaf showing venation, ×2. C holotype leafy shoot, ×0.5. D stoma from lower cuticle, ×250. F *P. thomasi* (pollen), ×500.

clearly established (Thomas and Bancroft 1913), most of the form-genera of leaves were easily distinguished. The unusual feature is that many reproductive structures are known in the form of rather large bisexual or unisexual flowers surrounded by protective bracts; they form the basis of useful extinct families which are grouped in the Benettitales.

Williamsoniaceae. Numerous species of the leaf-genera *Ptilophyllum* (fig. 8.9*A–C*), *Zamites*, *Otozamites* and probably *Pterophyllum*, that occur in the Yorkshire Bajocian strata, are associated with various unisexual flowers of *Weltrichia* (male) and of *Williamsonia* (female) (fig. 8.9*D*, *F*, *G*) some of which were even as much as 5–10 cm across when open. Seeds and pollen (fig. 8.9*E*) are known, but the habit of the plant is uncertain. Williamson's well-known reconstruction of '*Zamia*' *gigas*, which was probably based on his knowledge of the habit of Recent cycads, included a fossil stem *Bucklandia* which had not been shown to be directly associated with these leaves (*Zamites*) and flowers (*Weltrichia sol* and *Williamsonia gigas*). The later reconstruction by Sahni (1932), based on Indian petrified material, turns on a *Bucklandia* supposedly in association with a *Ptilophyllum* leaf. Except that all these stiff pinnate leaves are likely to have been arranged in a radiate crown, very little is known of the appearance of these plants (cf. Harris 1969, fig. 59*C*).

Wielandiellaceae. *Anomozamites* (segmented) and *Nilssoniopteris* (unsegmented lamina) (fig. 8.10) have pinnae or laminae borne mid-laterally on the rachis, unlike *Ptilophyllum* and *Otozamites* in which the attachment is more nearly adaxial. The small bisexual flowers *Williamsoniella coronata* (fig. 8.10*C*) have been shown to belong to the same plant as *Nilssoniopteris vittata*, and the reconstruction has relatively slender forking stems; the reconstruction is however really derived from *Wielandiella/Anomozamites* from the Swedish Triassic Rhaetian.

8.3.3 *Lanceolate leaves with anastomosing venation*

Sagenopteris (fig. 8.11*A–C*) is a petiolate leaf with two pairs of lanceolate leaflets borne together at the petiole apex, although, probably because of leaflet abscission, the whole leaf is seldom seen. Although there has been some confusion over recognition of closely related species, the attribution to the same (as yet un-named) plant of the microsporophyll *Caytonanthus* and the megasporophyll *Caytonia* (fig. 8.11*E–G*) are well established (Harris 1964) by cuticle comparison and by association. The stem of the plant is only known

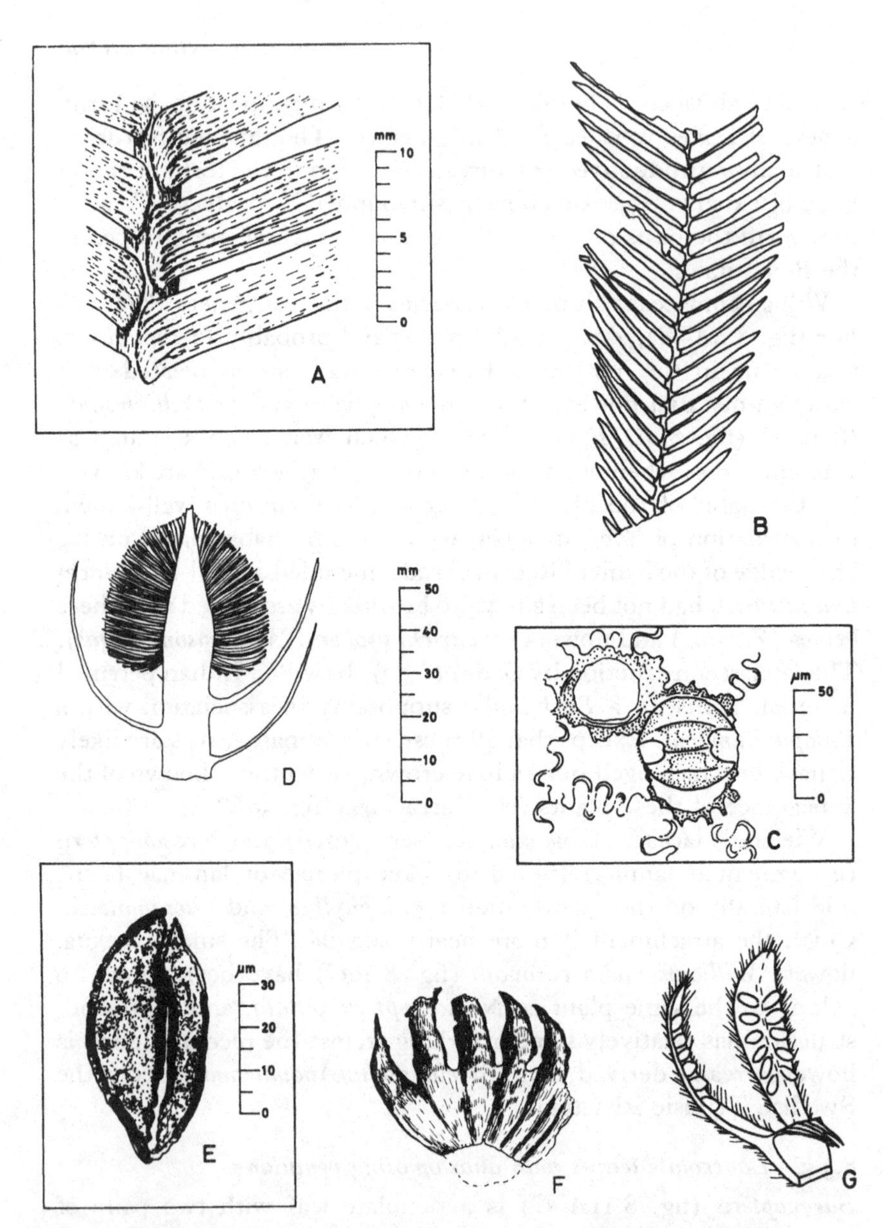

Figure 8.9. Jurassic Benettitales (Williamsoniaceae); Yorkshire, Bajocian. A–C *Ptilophyllum pectinoides* (Phillips) Morris; after Harris (1969). A detail of part of leaf, ×2. B lower part of large leaf, ×0.5. C cuticle detail, ×250. D *Williamsonia leckenbyi* Nathorst; restoration of fruit before loss of seeds, ×0.5. E *Monosulcites subgranulosus* Couper; dispersed pollen of *Weltrichia* type, ×500; after van Konijnenburg-van Cittert (1971). F *Weltrichia whitbiensis* (Nathorst), male flower, ×0.5. G *W. setosa* (Nathorst), restoration of part of flower, ×0.5; both after Harris (1969).

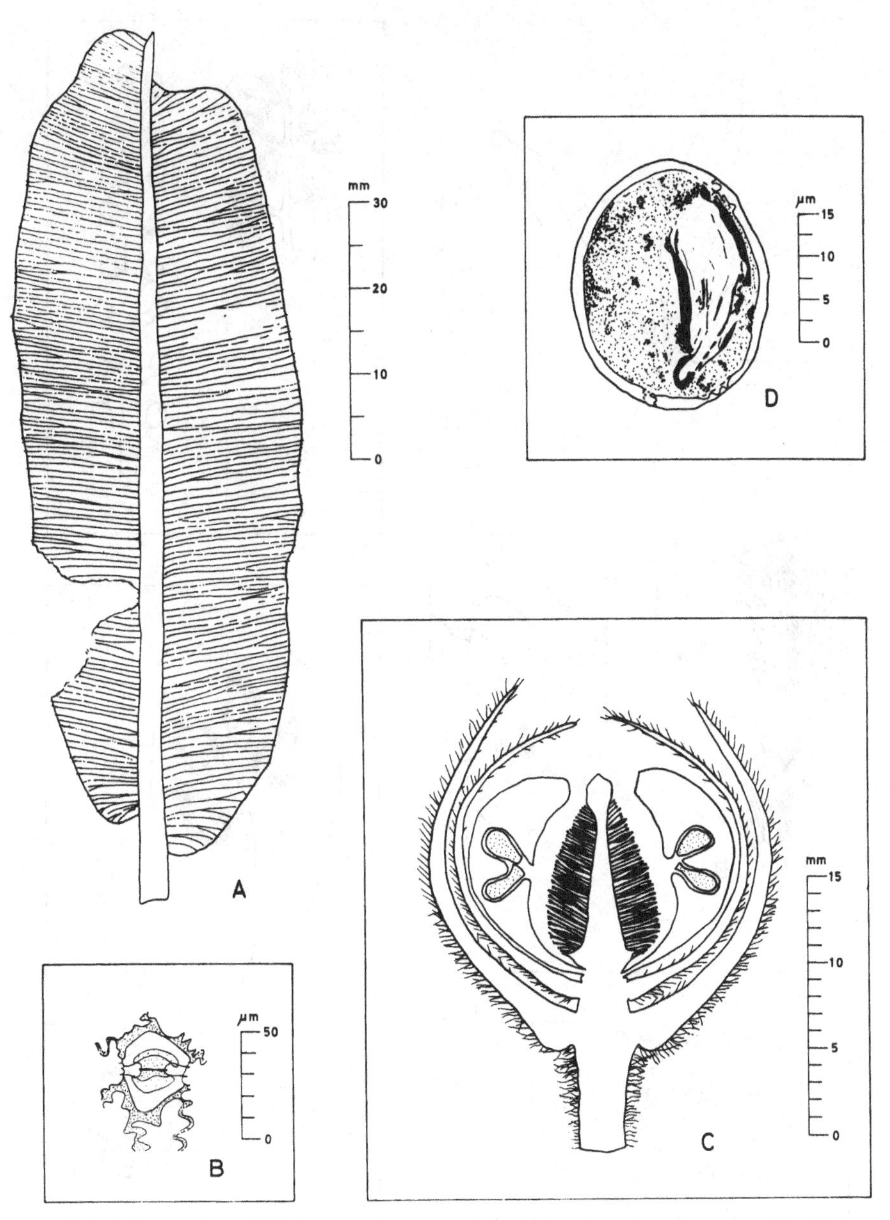

Figure 8.10. Jurassic Benettitales (Wielandiellaceae); Yorkshire, Bajocian. A *Nilssoniopteris major* (Lindley and Hutton), leaf ×1. B cuticle detail of same, ×250; both after Harris (1969). C *Williamsoniella coronata* Thomas; reconstruction of whole bisexual flower, ×2; after Harris (1969). D *W. coronata* (pollen), ×1000; after Couper (1958).

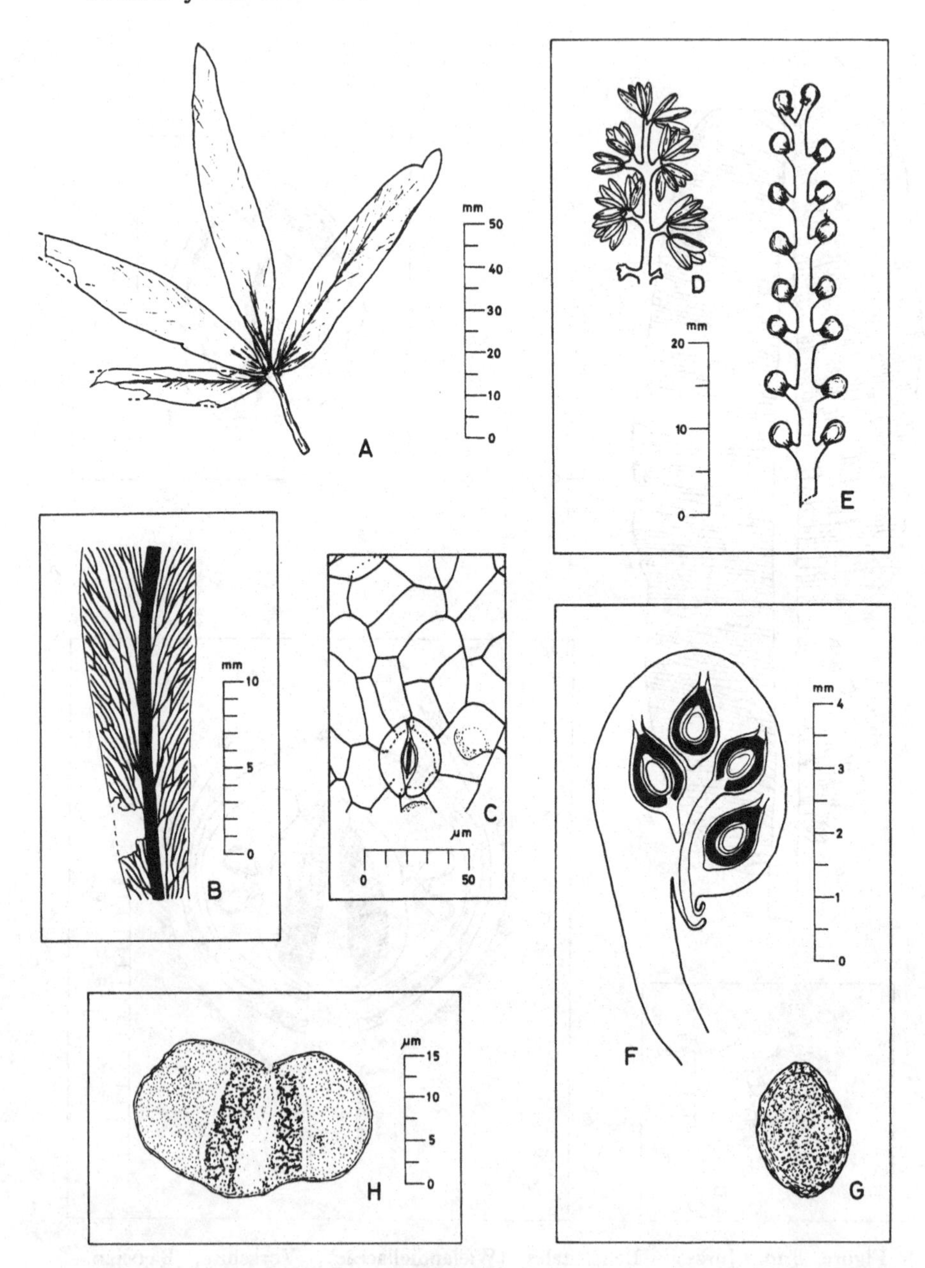

Figure 8.11. Jurassic Caytoniales; Yorkshire, Bajocian. A–C *Sagenopteris phillipsi* (Brongniart). A palmate leaf, ×0.5; after Thomas (1925). B venation of leaf, ×2; after Thomas (1925). C cuticle (lower), ×250; after Harris (1964). D *Caytonanthus arberi* (Thomas) Harris, reconstruction of part of male inflorescence, ×1; after

from a single small specimen. The dispersed pollen *Vitreisporites*, which occurs widely in Mesozoic assemblages, is closely similar to *Caytonanthus* pollen (fig. 8.11*H*) which is also found consistently in the micropyles of *Caytonia*. Delevoryas and Gould (1971) described *Perezlaria* from the Jurassic of Mexico; fructification and leaves may belong in this group.

8.3.4 *Fossils of Jurassic age but not represented in the Yorkshire Bajocian*

Only those fossils that can reasonably be thought to have existed in Bajocian time are considered. Many Triassic forms still important in Rhaetian time, such as the Peltaspermaceae, *Harrisothecium*, *Palaeotaxus* were probably extinct, and such forms as *Cycadeoidea* had not yet arisen.

'*Ephedra*' *chinleana* was redescribed by Scott (1960) from pollen of late Triassic age from Arizona. The next record of pollen is Cretaceous Barremian of western Europe, and the first megafossil is Cenozoic. Nothing is known of the Jurassic history of this group; it is possible that the Triassic pollen did not belong to the Ephedraceae.

Nipaniophyllum, *Carnoconites* etc., forming the Pentoxylales, (Sahni 1948) came from rocks in northern India which might be of Bajocian age but are probably later. The pollen grains, from the microsporophyll *Sahnia*, are monosulcate as are those of the Nilssoniaceae, but other features of the plant (see Sporne 1965) are diverse and unlike those of any other group. Some further fossils have more recently been found in New Zealand (Harris 1962).

8.3.5 *Miscellaneous unrelated fossils in the flora*

There are many indications that the systematic study possibilities of the rich Yorkshire Bajocian flora are by no means exhausted even with current methods; it is perhaps more realistic to say that in such a well-preserved large flora from a large outcrop area they will probably outlast human interest in palaeobotany.

Legend to figure 11 (*contd*)

Thomas (1925). E, F *Caytonia nathorsti* (Thomas) Harris. E reconstruction of megasporophyll, ×1; after Thomas (1925). F restoration of fruit in longitudinal section, ×7.5; after Harris (1933). G *Amphorispermum pullum* Harris, holotype, ×7.5; after Harris (1964). H *Caytonanthus* pollen, closely associated with *Sagenopteris*, ×1000; after Harris (1964).

Some examples are: *Amphorispermum pullum* (see Harris 1964) which is a widely dispersed seed resembling that of *Caytonia* but which does not have *Vitreisporites* pollen in the micropyle and does not agree in distribution with any *Caytonia* species; *Wonnacottia crispa* Harris (1942) which is a microsporophyll of unexpected form with a cuticle resembling *Anomozamites nilssoni*, presumed to be of the *Wielandiellaceae*; and the dispersed pollen *Eucommiidites* cf. *troedssonii* which occurs in almost all Yorkshire Bajocian pollen assemblages and is now recorded from the cone *Hastystrobus muirii* by van Konijnenburg-van Cittert (1971) but is still of unknown affinity (see chapter 9).

8.4 RECONCILIATION OF THE PALYNOLOGIC RECORD

The palynologic record for the Yorkshire Deltaic beds (Bajocian) was described in some detail by Couper (1958) and has since been re-examined by Muir (1964) and van Konijnenburg-van Cittert (1971). All the plant groups mentioned in sections 8.2 and 8.3 above have either known or probable representatives in the dispersed miospores, but in several cases there is some doubt or distinct incongruity of abundance between leaves and microfossils; there is also a short list of unallocated dispersed grains (see also table 9.1).

Couper did not allocate *Monosulcites minimus* between the *Ginkgoites* group and the Nilssoniales; and fossil ginkgophyte pollen evidence was very meagre indeed and Harris (1948), who described the only relevant grains, was very cautious. *M. minimus* is fairly common in the Gristhorpe beds (Middle Deltaic) and the Upper Deltaics, the *Ginkgoites* group megafossils being particularly common in the latter. The Wielandiellaceae (*Williamsoniella*) also produced grains of this type. Fortunately some of the Williamsoniaceae such as *Weltrichia spectabilis* are reliably associated with the larger *M. carpentieri* which is distinct; although obvious characters in such grains are few, progress could probably be made if any evidence solely from Recent pollen was firmly omitted. The unallocated *Eucommiidites* which is common in the Gristhorpe beds is essentially monosulcate as well as zonosulcate (Hughes 1961*b*) or irregularly tricolpate (van Konijnenburg-van Cittert 1971), and should be considered with these groups; it enters the succession in the Early Jurassic (Hettangian).

Classopollis occurs throughout the Yorkshire Bajocian, but not in the great abundance it shows in the Purbeck beds of England (earliest Cretaceous), and it is rare in the Upper Deltaic beds. This type of pollen enters the succession in the Hettangian as the pollen of *Hirmeriella*, but in the Bajocian it represents some species of *Brachyphyllum*. Although some new Jurassic species of *Classopollis* have been erected for North African material (Reyre 1970), fine stratigraphic distinctions within the genus have been difficult to make in the Jurassic because of very variable preservation.

Bisaccate pollen with distal saccus attachment has proved very difficult to handle taxonomically and remains a challenge to palynologists of the Mesozoic. Unlike the relatively featureless monosulcate grains which will almost certainly yield more characters on electron micrographs the saccate problem is one for great patience and discipline in optical microscopy. The grains described under *Abietinaepollenites* by Couper (1958), and under *Alisporites* by others, occurred commonly throughout the Jurassic but have not been attributed except to irrelevant Recent conifer groups. *Parvisaccites enigmatus* Couper 1958 has been properly distinguished but has also not been attributed.

8.5 JURASSIC PLANT CHARACTERS

Certain characters transgress in their degrees of advancement the limits of taxa of any classification. Presumably they indicate certain facts about the flora as a whole that if properly expressed are as much events in geological history as are the more traditional appearances of new taxa. It is not essential that their significance be understood before they are used in stratigraphy. It is preferable to avoid any evolutionary jargon, such as 'trends' or 'parallel evolution', in describing them.

8.5.1 *Pollen*

Mesozoic innovations in pollen structure did not survive through mid-Cretaceous time with any degree of importance, nor do they appear to have given rise to further developments. Presumably this was due to the unusual and overwhelming success of the tricolpate solution to the problem of random but unchangeable orientation of pollen grains after alighting on an adhesive stigmatic surface; this solution clearly marked the mid-Cretaceous rise of angiospermy. The Jurassic marginally-distal zonosulcate condition of *Classopollis* may have been connected with the same problem although rather more probably with some peculiarity of a drop mechanism than with any stigmatic surface; this applies also to the possibly proximal zonosulcate condition in *Eucommiidites* (although with that interpretation the tetrad orientation is uncertain; see Hughes 1961*b*). The plants concerned, *Hirmeriella* and *Brachyphyllum* for *Classopollis*, and an ovule with a very long micropyle for *Eucommiidites*, were clearly gymnospermous but were 'advanced' in a way not yet fully understood. The persistent occurrences of *Classopollis* in tetrads are believed to have been even more widespread than has been recorded; the standard palynological preparation method probably tends to break up tetrads. The function of such an 'experimental arrangement' has not yet been deduced beyond the thought that wind transport of pollen can hardly have been the main object.

No Jurassic pollen shows positive sculpture on the exine approaching the tectate condition; this first appeared in the Cretaceous (Barremian onwards) *Clavatipollenites* (see Kemp 1968). Such a device for adhesion in groups and/or to potential pollinators (insect and others) was developed strongly in miospores of the *Cicatricosisporites* and other groups, to some extent in the latest Jurassic but particularly in the early Cretaceous. Because the spore-bearing plants concerned were all herbs or scramblers, Coleoptera may have been the important agent as has been suggested on other grounds. Unless some unattributed miospores prove to have been pollen, the Jurassic gymnospermous plants appear to have established no direct correlation of pollen with insects. The elaborate infra-reticulum of *Classopollis* (Pettitt and Chaloner 1964) can surely only have been concerned with the shape and rigidity of the grain, as with the infra-reticulum in saccate grains.

The development of numerous small sacci observed in *Cerebripollenites* and in some species of *Tsugaepollenites* (al. *Applanopsis*) was a Jurassic character but with no continuing importance except for its survival in some *Tsuga* species.

There was little optically observable variety in Jurassic monosulcate grains except for the production of some very large elongate grains of *Monosulcites carpentieri*; these were associated with certain flowers of the *Williamsonia* type but no special function of this shape has been suggested. Unlike the other pollen referred to in this section, they remained rare.

8.5.2 *Flowers and seeds*

Although there is as yet no way to discover whether a fossil flower was separately 'coloured' from the plant, it seems probable that it was so, at least in part. It seems unlikely that the attraction was entirely odour rather than a visual one, because for an open actinomorphic form of flower it would hardly be necessary to develop the use of an odour. The gymnospermous flowers of *Williamsonia* and allied plants were relatively large, up to 10 cm across, and they probably represented the results of the first change-over from concealing the ovule as in many Carboniferous pteridoperms, to advertising the late Triassic flower (*Sturianthus*) and then protecting its seed from digestion. The assumption that these first flowers were concerned with distribution rather than pollination is supported by lack of specialisation of the associated *Monosulcites* pollen. From the size of the flowers it is reasonable to look for a reptile rather than an arthropod, and presumably some smaller specialised herbivore rather than a bulk vegetation feeder. Flying reptiles and birds probably did not practise any appropriate subtleties of flight; their earliest fossils are as yet only from the late Jurassic. In the late Triassic when these flowers were first widespread, herbivorous reptiles included the therapsid *Dicynodon* group; by Jurassic and particularly by early Cretaceous times there were more herbivores of different groups such as the Ornithischia (see chapter 6).

Little is known of the fully developed *Williamsonia* fruit and as Harris (1954) has pointed out the seeds so far described are undeveloped specimens from undeveloped fruit, and in most cases the mature and dispersed seed has not even been recognised; the main cutinised layer appears to have been the nucellus with doubtless some kind of stone layer in the late integument outside. As described by Harris (1961), *Beania* seeds had a compact woody layer corresponding to a stone overlain by a fleshy outer integument, and the megaspore membrane was heavily cutinised as is standard in living gymnosperms; but these seeds were fully developed.

8.5.3 *Leaves and stems*

Mid-Jurassic gymnospermous leaves were remarkably uncomplicated, being confined to simple or subparallel venation with the minor exceptions of anastomosing venation in *Ctenis* and *Sagenopteris*. However, pteridophytes of the same age, such as *Clathropteris*, had developed a large lamina with a kind of reticulate venation as also had enigmatic late Triassic *Furcula*, and the Jurassic/Cretaceous *Hausmannia*.

Cuticles of Mesozoic gymnospermous plants were probably

thicker than those of any other dominant group before or after, but perhaps this was more connected with the maintenance of leaf-shape rigidity than with any directly xeromorphic characters. Palaeozoic trees, as opposed to shrubs, had only achieved a genuine leaf lamina with parallel (*Cordaites*) or subparallel dichotomous (*Psygmophyllum*) venation; *Glossopteris*, which may have been truly arborescent, used anastomosing venation. It seems possible that as development of a viable leaf lamina for tall trees was not achieved until mid-Cretaceous time, cuticle diversity became important and is indeed still a factor in gymnospermous tree distribution at the present day. It could be argued that there is no direct evidence of tall trees in British mid-Jurassic fossils, but elsewhere, as in the Triassic of Arizona and in the British Cretaceous/Barremian of the Isle of Wight, the evidence for them is clear.

The many still pinnate leaves of *Ptilophyllum*, *Zamites* and others were presumably part of crowns on unbranched manoxylic stems. The stems *Bucklandia*, believed to belong to plants bearing *Williamsonia* flowers but certainly associated with *Ptilophyllum* leaves, were not more than 5–10 cm in diameter, but were covered with appropriate leaf-base scars. The time-distributions of the species of *Bucklandia* and of the Williamsoniaceae do not agree well, although poor preservation of stems in general, when no petrifactions are available, may account for this. To conclude, although most reconstructions emphasise cycad-like plants (fig. 8.12), they must in the Mesozoic have been relatively small and inconspicuous like *Cycas rumphianus* in present-day coastal Malaysia. Any large trees must have been the 'conifers' but possibly even they were not of any great height; our reconstructions may all subconsciously rely on Carboniferous or Recent tree dimensions.

8.5.4 *Roots*

Almost nothing is known of the evolution of deep true-root systems of trees. However, although such roots are familiar in present-day vegetation, they are only characteristic of plants which are either seeking a relatively distant water-table or spreading their catchment widely in a deep unsaturated soil layer. In fossil floras, which in pre-Cenozoic cases are entirely from aggradational land with by definition a high water-table, such roots probably did not occur.

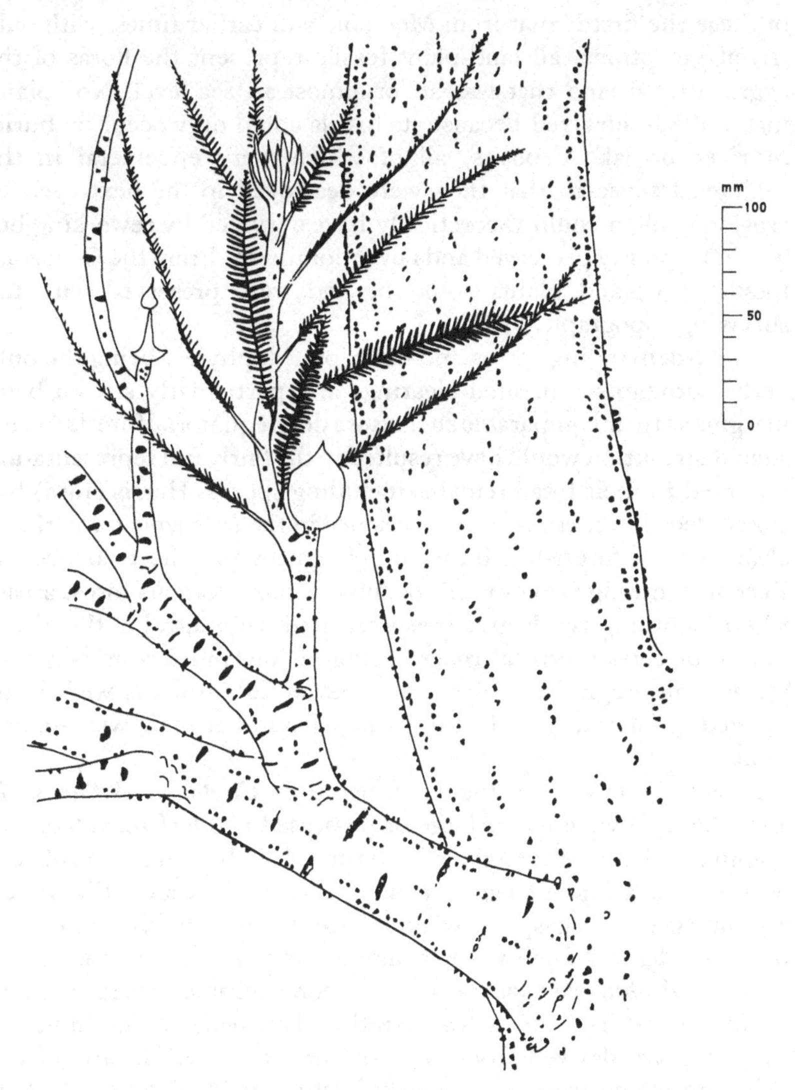

Figure 8.12. Reconstruction of a plant composed of axes of *Bucklandia pustulosa* Harris, flower *Williamsonia leckenbyi* and leaf *Ptilophyllum pecten*, ×0.25; after Harris (1969).

8.6 PALAEOECOLOGIC DISCUSSION

Although each section of this chapter is to some extent speculative, the assumptions made are not necessarily cumulative or interdependent. They should however be labelled as assumptions and of these the first is that from Mesozoic and earlier times, with only trivial exceptions, all land-plant fossils represent the floras of the aggradational land that was at, or almost at, sea level. No upland material has survived because its fossils could only occur by burial in river or lake deposits, all of which were ephemeral in the geological sense in that they were destroyed in the next cycle of erosion. Pollen could theoretically have survived by reworking but would be poorly preserved and never dominant. From the Cenozoic, fossils of upland plants could be and were preserved with the surviving topography.

Light-demanding forms and other pteridophytes, being the only herbs, probably inhabited clearings and particularly stream bank margins as their comparable successors do in equatorial forests today; such distribution would have resulted in the fairly full representation observed in their fossil remains including spores. Harris (1958) has concluded that there is ample evidence in the Mesozoic for periodic clearance by forest-fire initiated by lightning. The evidence for Recent lightning strikes is all for hilly or mountainous forest areas, where lightning is selective presumably under control of the distribution of certain crustal rocks; firing of lowland forests is not so certain, although lowland occurrences of fulgurites is well documented (Harland and Hacker 1966) from Permian wind-blown sands.

There is no evidence that such plants as *Caytonia*, *Williamsonia* and *Nilssonia* were either large or possessed of a leaf-form likely to dominate all other vegetation; from their distribution they probably occupied clearings of lagoonal or back-swamp origin; the varied assemblages of megaspores which had a scattered distribution, were probably derived from water plants occurring in these same areas. *Baiera* and *Ginkgoites* leaf remains are concentrated in certain sandy channel deposits (such as the Whitby plant bed), which indicates that they are derived from exposure and removal of parts of the delta-top which were not normally disturbed. Any deduction that they were of upland origin, or even that the plants were fully developed trees, is not warranted on the evidence of the leaves alone.

The above picture does not, however, immediately account for the palynological assemblages of the Yorkshire Deltaic beds originally described by Couper (1958) and subsequently studied in more detail in their sedimentary context by Muir (1964, unpublished). Local abundance in certain sediments of such forms as *Classopollis*, *Exesipollenites*, and *Abietinaepollenites* was interpreted by Muir as representing hinterland conifers; although not the basis of her argument, this whole idea of the importance of upland (hinterland) floral elements is connected with the supposed responsibility of marine 'transgression' for such assemblages (Chaloner 1958, Kedves 1960). Hughes and Moody-Stuart (1967*a*) working on the admittedly different geological conditions of the Early Cretaceous British Wealden found similar kinds of abundance, also apparently unsupported by megafossils, but concluded that Wealden *Classopollis* could well have represented a 'mangrove' sea-margin floral element; this conclusion is based on the details of the palynologic assemblage and lithology of the sample. Jung (1974) deduced on anatomical grounds that the late Jurassic *Brachyphyllum* conifer was a sea-margin halophyte. The bisaccate *Abietinaepollenites*, from the facts that abundance was superimposed on traces of other recognisable assemblages and that the samples involved were of varying lithology, was deduced to have belonged to plants widespread throughout the delta; the concentrations of pollen were believed to have been due to the size, flotation and other properties of the grains. In terms of the megafossils *Classopollis* appears to have been correctly attributed to the gymnospermous Brachyphylls, and the extension of some of these lowland forest dominants over the actual sea-margin seems reasonably likely. The attribution of the saccate grains with distal saccus attachment is not known, and is presumably for the present submerged in the gymnospermous Linearphylls. Obviously the discovery of new details about any of the plants would quickly change the speculative palynological statement, possibly back towards Muir's contention.

8.7 FLORAL SUCCESSION

From a study of the very numerous individual plant beds that he has worked in the Yorkshire Bajocian, Harris (1953) has distinguished a floral succession of delta-flat colonisation by plants which began with pure *Equisetites* stands; these were normally succeeded by conifer and fern assemblages, and only somewhat rarely (as in the famous Gristhorpe Bed) by a fully developed flora from a fine mudstone lithology probably representing lagoons or abandoned channels in the delta. In these special cases fruits and leaves are often found together indicating close proximity of the locality to the original site of the growing plant; an alternative might be short distance water transport of large plant limbs.

Critical fossil evidence

It is interesting that this record indicates that the Linearphylls and Brachyphylls (conifers) were the main trees colonising the aggradational lowland, and suggests that benettites, cycads and caytonias etc. were more local elaborations. The conifer vegetation may well have presented the appearance more of a thicket than of a modern spruce forest, and it remains a serious difficulty to imagine and reconstruct a climax lowland forest of this kind without either angiosperms or even familiar gymnosperms.

Although some of these plants doubtless also existed in upland vegetation at the time, their lowland presence as indicated by megafossils and miospores alike was the important centre of evolution.

9 Early Cretaceous gymnosperms

Various prejudices have contributed subtly to a situation in which 'early Cretaceous gymnosperms' do not appear to have called for any appreciable separate recognition or study; indeed, they have all appeared to conform either to Jurassic or to Recent groupings without trouble. As a consequence, and disregarding the 'upland' and other untestable hypotheses, the immediate ancestors of the angiosperms of the mid-Cretaceous must have been among the normal lowland gymnosperms of the early Cretaceous, of which no general assessment is readily available.

There are some satisfactory published fossil records such as that of the Wealden flora of southern England (Seward 1894–5, 1913; Watson 1969), but within most of the recognised plant groups these fossil occurrences have not been considered independently of the better known mid-Jurassic material; the Mesozoic flora has been thought of as a whole without evolutionary change, even to the extent sometimes of denying that any change had occurred. It is also necessary, and perhaps a little more difficult, to emphasise distinctions of the Cretaceous fossils from their extant but (for this purpose) irrelevant successors. As far as possible the Berriasian and Valanginian fossils will be discussed in groups believed to have been descended from the Jurassic plants in the classification used in the previous chapter; the use of extant taxa of all low ranks will be avoided, a policy also advocated long ago by Berry (1911, p. 389) in discussing the Potomac flora.

The latitudinal distinctions of floral provinces appear to have been stronger in the Cretaceous than in the Jurassic. 'Ginkgophytes' occupied high northern latitudes, and '*Classopollis* brachyphyll conifers' low latitudes; between were saccate-pollen conifers which overlapped parts of the ranges of the others but were not important in the palaeotropics.

At the present stage of study every single gymnosperm group present in the early Cretaceous should be considered capable of having given rise to some angiospermid characters; the many botanical prejudices against accepting any one of them should be reconsidered on the basis of *fossil* evidence alone. The possibility of a hypothesis of polyphyletic origin of angiosperms leading to an entirely new classification of seed plants should not be excluded.

9.1 'GINKGOPHYTES'

The greatest development of the whole history of this group was in the late Jurassic and early Cretaceous of middle and especially of high northern palaeolatitudes, particularly in Asia where three families each with several genera have been recognised; the differences of cuticle structure and other detail are often more striking than those of gross leaf morphology. The whole group may then have been as large as any living conifer group now is, with the possible exception of the Cupressaceae. Ginkgophytes were also present in high southern palaeolatitudes of Argentina which may suggest progressive dispersal in Jurassic to Cretaceous time from the tropics to polar regions rather in the same way as Axelrod (1959) has suggested a dispersal for Cretaceous angiosperms.

It is important to emphasise that there is now only one isolated living species of ginkgophyte, which is well known and provides a powerful but probably distorted reconstruction-picture for such a group. For example, when it is recorded that recognisable ginkgoalean wood is only found from the Palaeogene onwards, this is quite possibly inaccurate because the wood of *Ginkgo biloba* L. (the only source of information) is highly evolved and thus almost certainly atypical for the group; earlier wood from the group could well be unidentifiable on this restricted information. It may be of course that this argument should apply also to reproductive structures, because it is on the whole unlikely that sixty million years of natural selection has failed to modify the morphology in the generations leading from the early Tertiary to this single relict taxon of very limited geographic distribution.

Krassilov (1970, 1972*a*) described the East Asian ginkgophytes in the three families Ginkgoaceae, Karkeniaceae and Pseudotorelliaceae; the leaves of these families were respectively fan-shaped, wedge-shaped and linear-lanceolate, and the cuticle details differed. The full distinction, however, is based on the female reproductive structures; in the fossil Ginkgoaceae (Jurassic and Cretaceous) there are numerous associated dispersed ovules, *Allicospermum* (fig. 9.1),

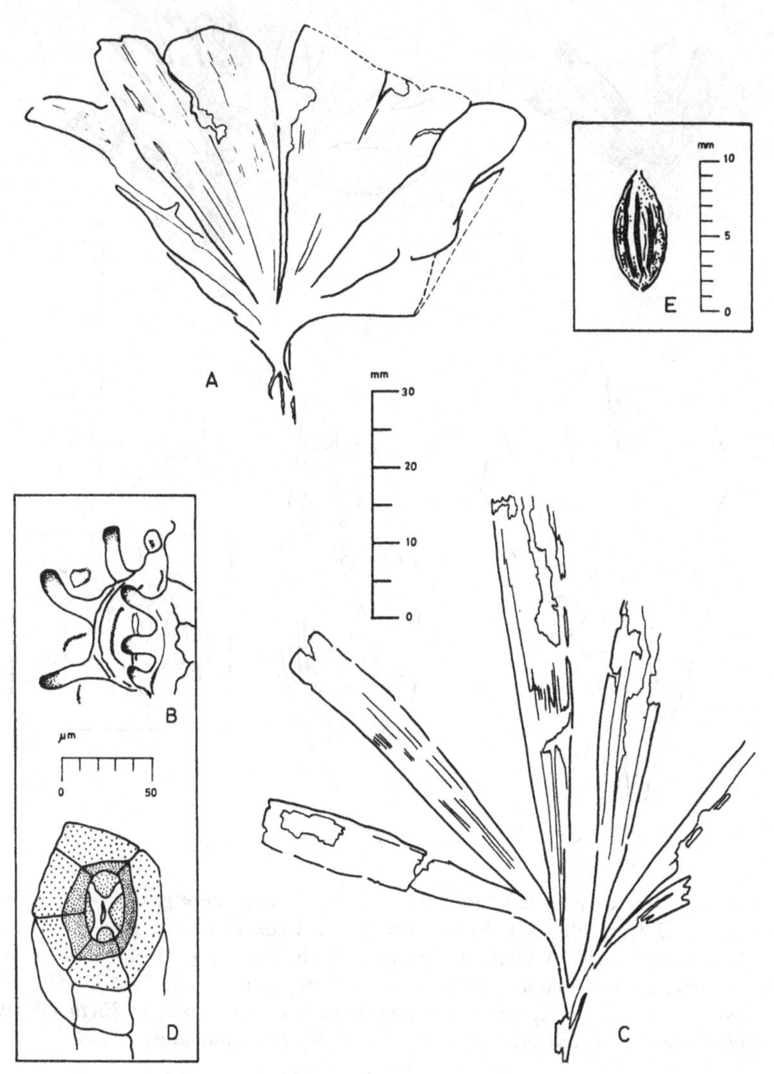

Figure 9.1. Cretaceous 'Ginkgophytes' of the Ginkgoaceae group; Bureja River area, Far Eastern USSR; after Krassilov (1972*a*). A *Ginkgoites longipilosus* Kr. 1972; early Cretaceous; ×1. B cuticle, ×250. C *Baiera manchurica* Yabe and Oishi (1913); earliest Cretaceous; ×1. D cuticle, ×250. E *Allicospermum adnicanicum* Kr. 1972; early Cretaceous; ×2.

but the detail of the megastrobilus unfortunately still relies only on that of the living *Ginkgo biloba*. It is even possible that the true Mesozoic fructification is so unlike *G. biloba* as to pass so far unrecognised in this context.

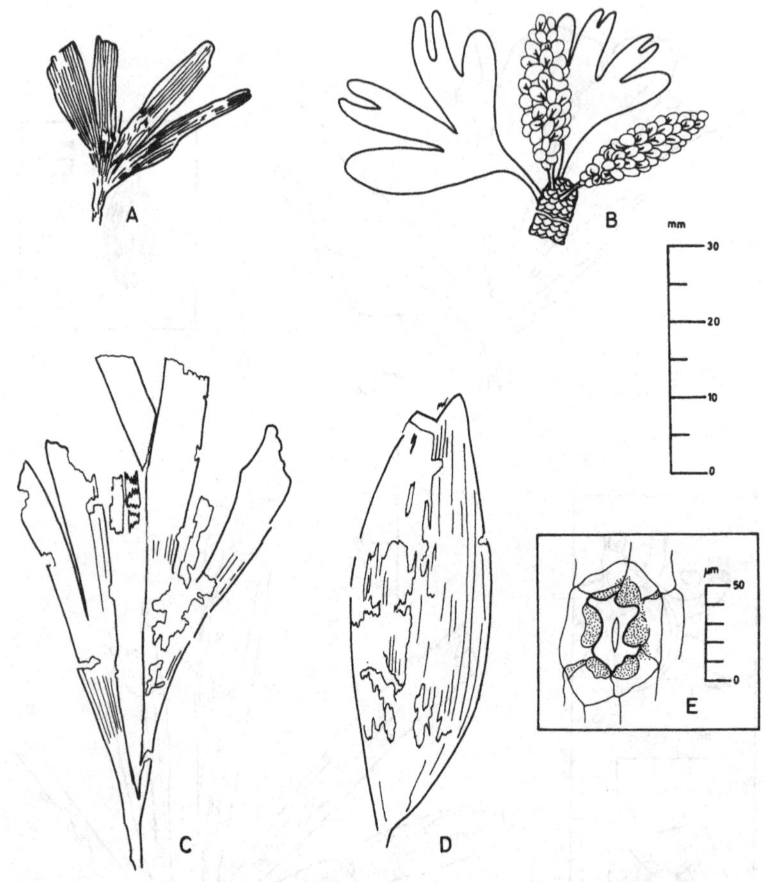

Figure 9.2. Cretaceous 'Ginkgophytes' of the Karkeniaceae group; early Cretaceous. A, B from Baquero Formation, Santa Cruz Province, Argentina; after Archangelsky (1965). A *Ginkgoites tigrensis* Arch. 1965; ×1. B *Karkenia incurva* Arch. 1965; reconstruction, ×1. C–E from Bureja River, Far East USSR, after Krassilov (1972*a*). C *Sphenobaiera umaltensis* Kr. 1972, ×1. D *Eretmophyllum glandulosum* (Samylina) Kr. 1972, ×1. E cuticle, *E. glandulosum*, ×250.

Karkenia (fig. 9.2) is a short axis with over a hundred closely packed small ovules each with a projecting micropyle and a short stalk; the family Karkeniaceae was known from both hemispheres and Krassilov (1970) considered the leaf *Sphenobaiera* to belong to the same plant. *Pseudotorellia* (fig. 9.3) is accompanied by the bract *Umaltolepis*, with cuticle similarities to the leaves, and by the dispersed oval seed *Burejospermum*; this group was confined to the northern hemisphere, but was very diverse in the large area from East Greenland to the Soviet Far East (Lundblad 1968).

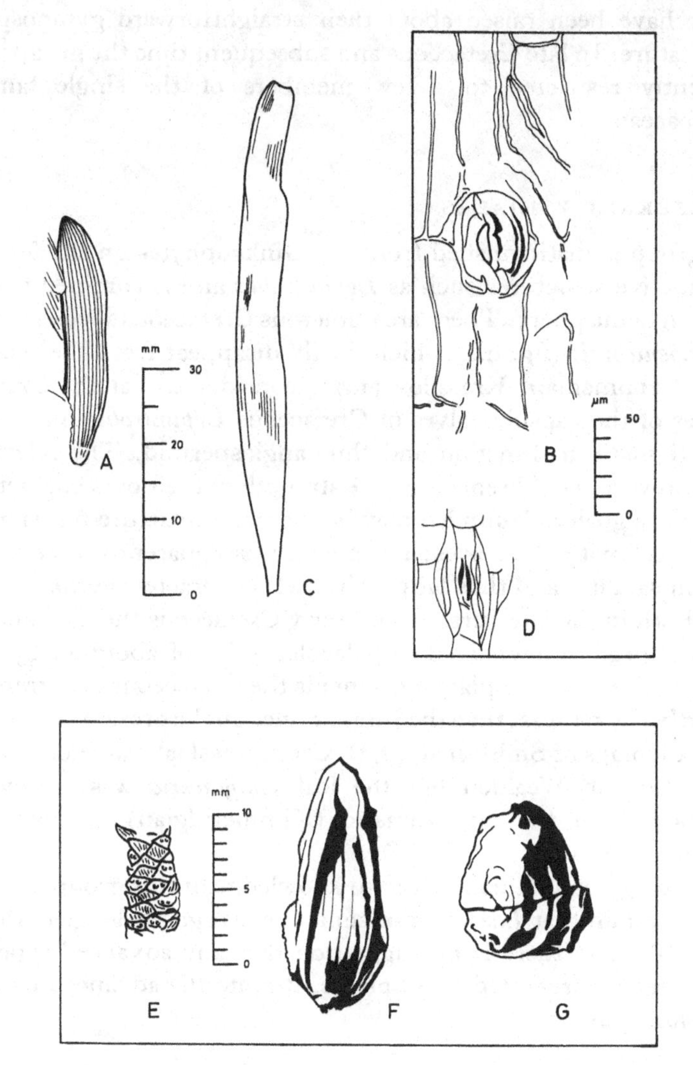

Figure 9.3. Cretaceous 'Ginkgophytes' of the Pseudotorelliaceae group; from the Bureja River, Far East USSR. A, B *Pseudotorellia ensiformis* (Heer) Doludenko, after Vakhrameev and Doludenko (1961). A leaf, ×1. B cuticle, ×250. C–G after Krassilov (1972*a*). C *P. angustifolia* Dol. 1961 emend. Kr.; leaf, ×1. D cuticle, ×250. E stem base, ×2. F *Umaltolepis vachrameevii* Kr. 1972, cone-scale ×2. G *Burejospermum crassitestum* Kr. 1972, seed ×2.

It is clear that there were several separate reproductive structures, and perhaps none of them resembled the living *Ginkgo* very closely. Considerable further study is necessary to classify them, but no doubts have been raised about their straightforward gymnospermous nature. In late Cretaceous and subsequent time the group was apparently restricted to a few members of the single family Ginkgoaceae.

9.2 CZEKANOWSKIALES

This group is distinguished from the Ginkgophytes on the female reproductive structures such as *Leptostrobus* and is confined to the northern hemisphere. There are numerous Cretaceous species of the leaf *Phoenicopsis* (fig. 9.4) which finally disappear from the record in the Cenomanian. Krassilov (1970) considered that the contact surfaces of the capsule valves of Cretaceous *Leptostrobus* may have been stigmatic in function and thus angiospermid. The relevant leaves however could remain on the strength of their own characters in the Ginkgoales. *Ixostrobus* may be the male structure but is only 'associated' with the leaves; the pollen was apparently very small and non-saccate, and thus lacking in useful characters at the optical level. Both in the late Jurassic and early Cretaceous the Czekanowskiales appear to have lived in palaeolatitudes of about 40–45° N; because of subsequent plate movements the Cretaceous occurrences fall further north-east than the Jurassic ones on the present continent (compare maps of Smith *et al.* 1973). Czekanowskiales are not known in the English Wealden but the leaf *Culgoweria* was originally described from the late Jurassic (Kimmeridgian) of northern Scotland.

Palynology cannot contribute to knowledge of this group until it is more certain that *Ixostrobus* or some other organ belongs to these plants. To date there is no suggestion that any advanced type of pollen grain represented these plants, despite the advanced nature of *Leptostrobus*.

9.3 CONIFERALES: PINACEAE

In most generally described early Cretaceous megafossil floras members of this family have not been recognised and listed (e.g. Samylina 1974), although the pine group must have become distinct at some time in the Cretaceous because of its rather full Cenozoic, and some minor late Cretaceous, records. It is necessary therefore to examine the few detailed discussions that have been published. Alvin (1953, 1960*b*) worked on the unusually well preserved Belgian

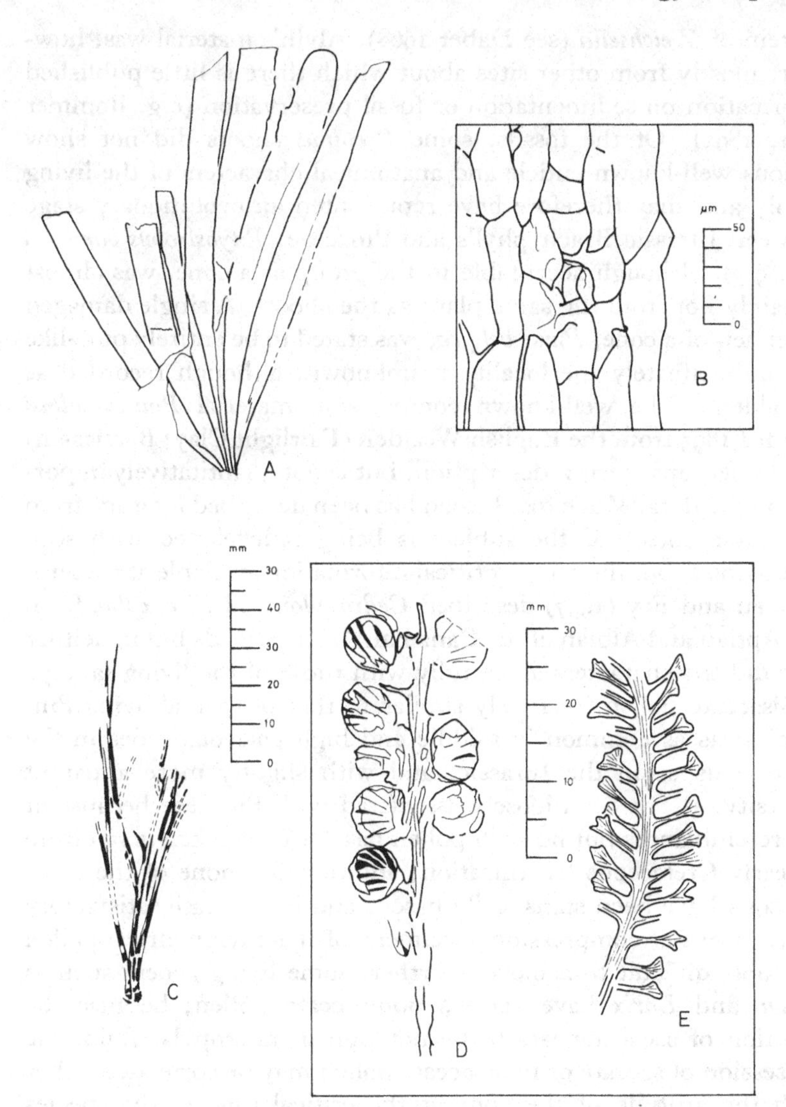

Figure 9.4. Early Cretaceous Czekanowskiales; from Bureja River, Far East USSR. A, B, D after Vakhrameev and Doludenko 1961. C, E after Krassilov (1972*a*). A *Phoenicopsis speciosa* Heer, leaf, ×0.5. B cuticle, ×250. C *Czekanowskia aciculata* Kr. 1972, leaf, ×0.5. D *Leptostrobus laxiflora* Heer, female fructification, ×1. E *Ixostrobus* ex gr. *heeri* Prynada 1951, male fructification, ×1.

Wealden from fissure deposits, the most famous of which at Bernissart (Seward 1900) is thought to be of approximately Hauterivian age because of the frequent occurrence (Anderson and Hughes 1964) therein of *Weichselia* (see Daber 1968). Alvin's material was, however, mostly from other sites about which there is little published information on sedimentation or fossil preservation (e.g. Bommer 1891, 1892). Of the fossils, some *Prepinus* shoots did not show various well-known cuticle and anatomical characters of the living family and may therefore have represented an evolutionary stage between Jurassic Brachyphylls and Pinaceae; *Pityostrobus bommeri* (fig. 9.5), although acceptable in the group as a cone, was almost certainly not from the same plant as the shoots. A single damaged specimen of a cone, *Pinus belgica*, was stated to be entirely pine-like but unfortunately the locality is unknown, although recorded as 'Wealden'. The well-known compression material *Pinites solmsi* Seward 1895 from the English Wealden (Fairlight Clay; Berriasian) would perhaps merit redescription, but is not quantitatively important in the flora. Much fossil wood has been described long ago from the Cretaceous and the subject is being redeveloped with scan microscopy, but there is no critical information available at present. Bannan and Fry (1957) described *Cedroxylon* and *Piceoxylon* from the Aptian and Albian of the Canadian Arctic Islands but in neither case did the characters agree fully with those of the living family.

Bisaccate pollen, relatively similar to that of several living Pinaceae, was as common in middle and high palaeolatitudes in the Cretaceous as in the Jurassic, and with slightly more apparent diversity. It has been loosely associated with Pinaceae because of the resemblances but no such pollen has yet been taken *in situ* from an early Cretaceous fructification; additionally, none of the comparisons have been statistically based, and in any case satisfactory descriptions of compression specimens of this asymmetrical pollen are most difficult to achieve. Further, some living genera such as *Tsuga* and *Larix* have various non-saccate pollen; because the function of sacci appears to be flotation in micropylar fluid, the possession of saccate or non-saccate pollen may be correlated either with the attitude of the cone at the critical time in the species concerned, or with a slow fertilisation and development schedule. The possible evolutionary origin of the bisaccate condition from monosulcate grains, and of the other saccate conditions from radially symmetrical grains, suggest that there is little scope for instant pollen attribution in this field.

Consolidation of knowledge of this family in early Cretaceous time seems possible but far from completion.

9.4 CONIFERALES: TAXODIACEAE

This primarily Tertiary family with a small number of scattered living relicts includes some accurately described cone material from the Belgian Wealden (Harris 1953*b*). *Sphenolepidium*, species of which (fig. 9.5) are common in the English Wealden (Berriasian) and in other floras, is often regarded as lying between the Jurassic Linearphylls (*Elatides*) and the Taxodiaceae. The Patuxent–Arundel (Aptian and Albian) part of the Potomac flora (Berry 1911), the Portuguese Buarcos (and earlier) floras and the Bohemian Cenomanian flora (see Knobloch 1971) include several species each of '*Sequoia*', in most cases with cones as well as twigs.

Most if not all the living representatives have simple pollen with a single distal protuberance. Although such pollen is common in mid-Tertiary deposits, it is not seen in early and mid-Cretaceous assemblages such as the Potomac (see Brenner 1963; Kemp 1970), and was obviously not a character of the group at these times.

9.5 OTHER CONIFERALES

Araucariaceae, Podocarpaceae and Cupressaceae have all been 'claimed' to occur in early Cretaceous floras, but always on the basis of individual organs and characters (often of pollen) or to meet the needs of evolutionary theory (Florin 1963). They should probably all be excluded, and the fossils concerned should be grouped for the present as 'early Cretaceous Brachyphylls or Linearphylls' or under some other neutral name.

In the southern hemisphere early Cretaceous conifers are attributed to Podocarpaceae and Araucariaceae for traditional reasons or in default of other apparent possibilities; such a mode of thought should not, however, be applied too rigidly before mid-Cretaceous time.

For the Podocarpaceae, trisaccate pollen grains or bisaccate grains with longer sacci than bodies are quoted as some of the living podocarp characters. The trisaccate grain occurs regularly in the Australian early Cretaceous (e.g. Dettmann 1963) and has been collected from the cone of *Trisacocladus tigrensis* from Patagonia (Archangelsky and Gamerro 1969), but is not recorded in the northern hemisphere as more than an isolated aberrant grain; the bisaccate grain with longer sacci is universal in occurrence but in assemblages such as those from the English Wealden it is always

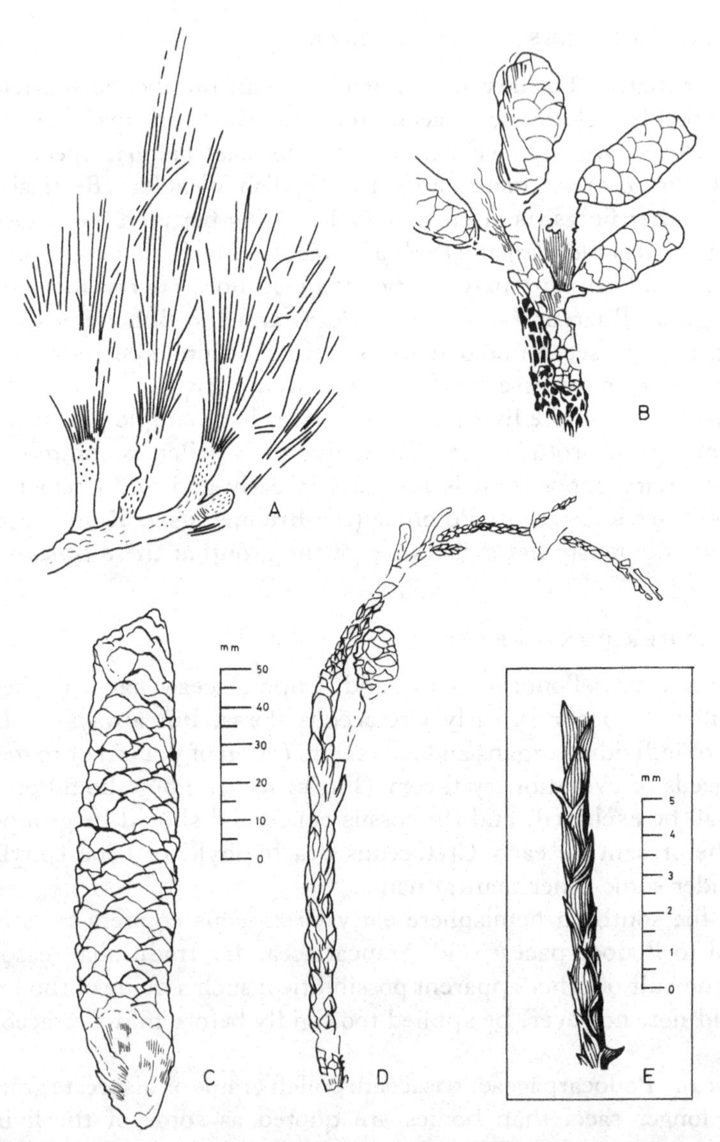

Figure 9.5. Early Cretaceous Coniferales; A–C, 'Pinaceae'; D, E, 'Taxodiaceae'. A *Pinites solmsi* Seward 1895, branch with leaf-bases and needles; Hastings Beds, Ecclesbourne, Sussex; after Seward, ×0.5. B *P. solmsi* branch with cones; same locality, after Seward, ×0.5. C *Pityostrobus bommeri* Alvin 1953, cone from Bernissart, Belgium; after Alvin, ×0.5. D *Sphenolepis kurriana* (Dunker) Schenk; leafy shoot, Hastings Beds, Ecclesbourne, Sussex; after Seward (1895), ×0.5. E *S. kurriana*, part of stem, Wealden, Belgium; after Harris (1953*b*), ×5.

markedly subordinate in numbers to the otherwise similar bisaccates with the normal shorter sacci suggesting that it may represent a regular variant grain despite its frequent recording as a separate taxon. The mainly European Barremian grain *Parvisaccites radiatus* Couper 1958 slightly resembles the pollen of some species of *Dacrydium* in the radial structure of the sacci. Fontaine (1889) described a *Nageiopsis* leaf species from the Potomac and Seward (1895) another from the English Wealden; *N. ussuriensis* was described from the early Cretaceous of Primorya. Little has subsequently been made of these but Krassilov (1974), when describing a Danian podocarp leaf with cuticle from the Bureja basin (Primorya), suggested that the cuticle of *Tritaenia* (*Abietites*) *linkii* (Roemer) Magdefrau et Rudolf from the German Wealden showed abaxial cuticle with three 'teilstreien' as in Podocarpaceae; this may prove to be a useful character in tracing an origin for the group.

Araucarites cone-scales in the early Cretaceous have been taken to represent the family Araucariaceae. The pollen grain named *Araucariacites australis* is universally as common in early Cretaceous as in Jurassic rocks; its Jurassic attribution however is to a *Brachyphyllum* species, and other Jurassic *Brachyphyllum* species have cones carrying *Classopollis* pollen. *Tomaxiella biformis* from the early Cretaceous of Patagonia has a male cone comparable with *Hirmeriella* of the Cheirolepidaceae and carries *Classopollis* (see Archangelsky 1968). *Classopollis* grains of separate distinct species were so common in the early Cretaceous of lower latitudes that the parent plants must have been dominants in some equatorial lowlands. This position appears to continue until mid-Cretaceous time when angiosperms take over equatorially, and it seems more logical to look from that time onwards for Araucariaceae as a group quite separately from the records of Jurassic and early Cretaceous Brachyphylls. The earliest Cretaceous *Classopollis* in western Europe is particularly found adhering together in tetrads even through the process of dispersed miospore preparation; this is a pollen character which may prove to be in advance of normal gymnosperm design in that it appears to provide for a symmetrical distribution of potential germinal apertures.

Cupressaceae cannot be distinguished through comparison with the very simple characters of the pollen grains of living members. The shoots and leaves, however, would have been compact and durable in fossilisation. *Frenelopsis* Schenk 1869 from the German early Cretaceous (Hauterivian–Barremian) was named after the living Australian genus *Frenela*, although Heer at about the same time regarded the fossil as an *Ephedra*; it is known also from the Potomac group and from Portugal, both from Barremian time onwards; it is not known from the main English Wealden flora which is of Berriasian age, although some shoots and cuticle under the name *Sciadopitytes* have recently been recorded there (Oldham in press). *Widdringtonites* Endlicher, originally described from the Palaeogene, has been recorded from the Patapsco (upper Potomac) and even from late Jurassic of Karatau, Kazakhstan, by Turutanova-Ketova (1930). Appropriate fructifications have not yet been described from any early Cretaceous flora.

9.6 TAXACEAE

This group has been distinguished from other linear-leaf gymnosperms in Jurassic time by the fructifications, both Jurassic and Recent, and by the leaf cuticles (Florin 1958). The only early Cretaceous fossils appear to be the seed *Vesquia* from the Belgian Wealden (Alvin 1960*a*) and some leaves *Taxocladus obtusifolia* Prynada from Primorya. Dispersed cuticle studies appear to offer some hope of clarifying the evolution of this group of plants providing robust cuticle material.

9.7 CHLAMYDOSPERMIDAE

The three living genera *Ephedra*, *Gnetum* and *Welwitschia* are so different from each other that it is difficult to see that any benefit has been derived from grouping them; however, only the Ephedraceae are considered here for a possible early Cretaceous presence on the basis of the multicolpate pollen *Ephedripites* found in most successions from Barremian time onwards, and *Steevesipollenites* from the Cenomanian onwards.

The late Triassic *Equisetosporites* has been regarded as '*Ephedra*' pollen by Scott (1960), but it has not been found regularly later than the Triassic and there is no other appropriate record to consider until the Barremian pollen mentioned above. Leaves and stem fossils are not known before the late Tertiary, although they were apparently very suitable for fossil preservation; the 33 living species are all found in semi-arid habitats in which the fossilisation potential is automatically very low. Antecedents of these living forms may well have been living in the aggradational land vegetation but if they were the leaf

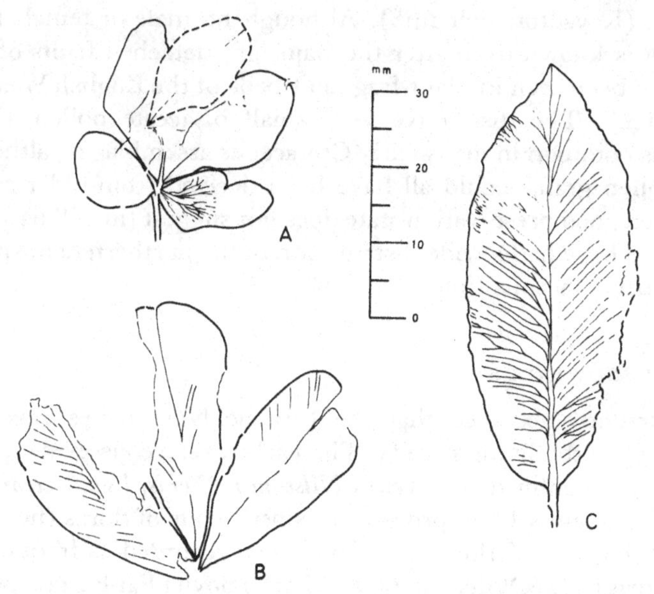

Figure 9.6. Early Cretaceous Caytoniales. A *Sagenopteris mantelli* (Dunker), leaf from Hastings Beds, Ecclesbourne, Sussex; after Seward (1894), ×1. B *S. mantelli*, leaf from Suifun basin, Far East USSR, Aptian age; after Krassilov (1967), ×1. C *Caytonia orientalis* Kr., leaf from same beds, ×1.

characters would probably have been very different from those of the arid-land plants. The suggestion (Hughes 1961*b*) that the early Cretaceous *Spermatites* ovule, with *Eucommiidites* pollen in the pollen chamber, belonged to this group is not very strong; the *Spermatites* micropyle was very long as is the case in *Ephedra*, but this fossil pollen grain cannot be connected with any of the pollen of known living species.

Although the only positive evidence here is from dispersed miospores, there is a strong suspicion that some as yet unrecognised Mesozoic plant group gave rise to a line now represented by the *Ephedra* relicts; such a group might prove not to be classifiable as 'conifer' or 'cycadophyte' or even as an ephedroid in the current sense.

9.8 CAYTONIALES

The leaf *Sagenopteris* (fig. 9.6) has been recorded from the English Wealden (Berriasian) and from early Cretaceous of the USSR (Krassilov 1967); it is also reported from Czechoslovakia and from

Sakhalin (Kryshtofovich 1918). Although no male or female fructification is known from after the Bajocian, detached fruits of this type have been seen in plant-fragment beds of the English Wealden (Berriasian). The distinctive very small bisaccate pollen *Vitreisporites* is common in many early Cretaceous assemblages; although these pollen grains could all have been derived from older rocks, their often good preservation state does not suggest this. This group appears to have had a wide distribution in the northern hemisphere until mid-Cretaceous time.

9.9 NILSSONIALES

It is currently accepted that the Jurassic Nilssoniales gave rise eventually to the living cycads. The early Cretaceous evidence for this group is confined to leaves of *Nilssonia*, *Ctenis*, *Pseudoctenis* and others which are seldom prominent constituents of floras (fig. 9.7), although cuticles of this group have been recorded as frequent in plant debris beds (Oldham in press). In the Soviet Far East (Siberian province) distinctive leaves such as *Heilungia* and *Aldania* (fig. 9.7) have unusual venation and pinnule margins, and are more prominent than *Nilssonia*, some species of which also have dentate margins. There is as yet no record from any area of a male or female cone and perhaps the only current chance of such evidence is from plant debris beds. In the very late Jurassic Morrison formation of Utah there are numerous but crude petrifactions of 'cycadophyte' fruits (Chandler 1966; Scott 1969) of several unrelated unknown types. Monosulcate pollen certainly existed in the early Cretaceous but it must be shared with some other plant groups; it is possible that scan microscopy may produce enough characters to enable firmer attributions to be made.

Nilssonia was the only one of these leaf types to persist into the late Cretaceous; the others disappearing from the record by Aptian time appear to be an unsatisfactory link between Mesozoic abundance and the minor diversity of living cycads. There is thus something disturbingly unsatisfactory about this link-story because the leaves are common and if their fructification resembled the Jurassic *Beania* it should have been recognised and collected.

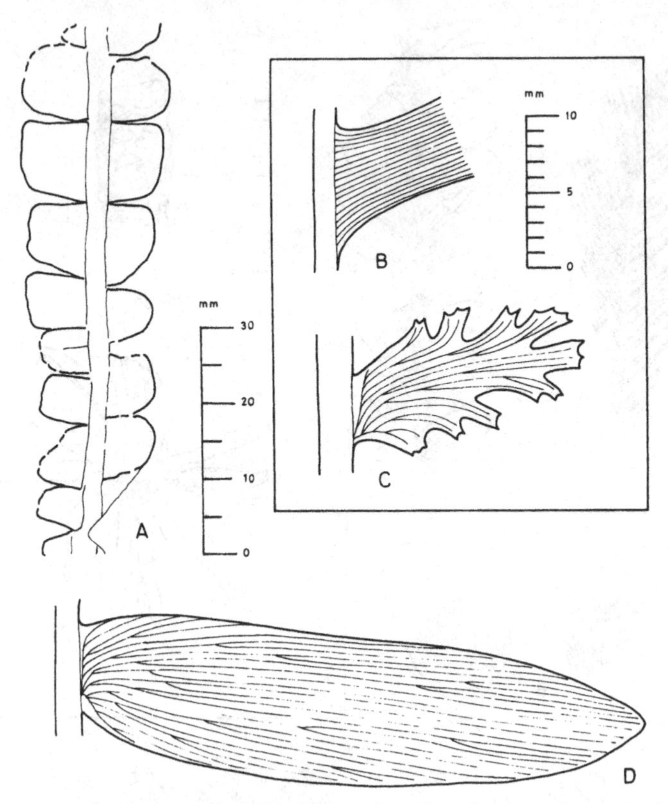

Figure 9.7. Early Cretaceous Nilssoniales. A *Nilssonia schaumbergensis* (Dunker) Nathorst, leaf after Krassilov (1967), ×1; from south-east Primoria, Far East USSR; Valanginian age. B–D figures after Vakhrameev (1968). B *Pseudoctenis* sp., leaf detail, ×2. C *Aldania umanskii* Vakhr. and Lebedev, ×2. D *Heilungia aldanensis* Samyl., ×1.

9.10 BENETTITALES

This term is used in the normal European sense to include Williamsoniaceae, Wielandiellaceae and Cycadeoidaceae; the American usage of the last of these names at ordinal level (e.g. Delevoryas 1962, p. 128) is considered to be unnecessarily confusing (see glossary) when the higher levels of the hierarchy have so little objective meaning and are frequently changed in rank.

The early Cretaceous Williamsoniaceae are mainly represented by leaves and cuticles of *Zamiophyllum*, *Otozamites* (fig. 9.8), *Dictyozamites* and others, which are virtually all absent from late Cretaceous floras. The small number of flowers which have been recorded

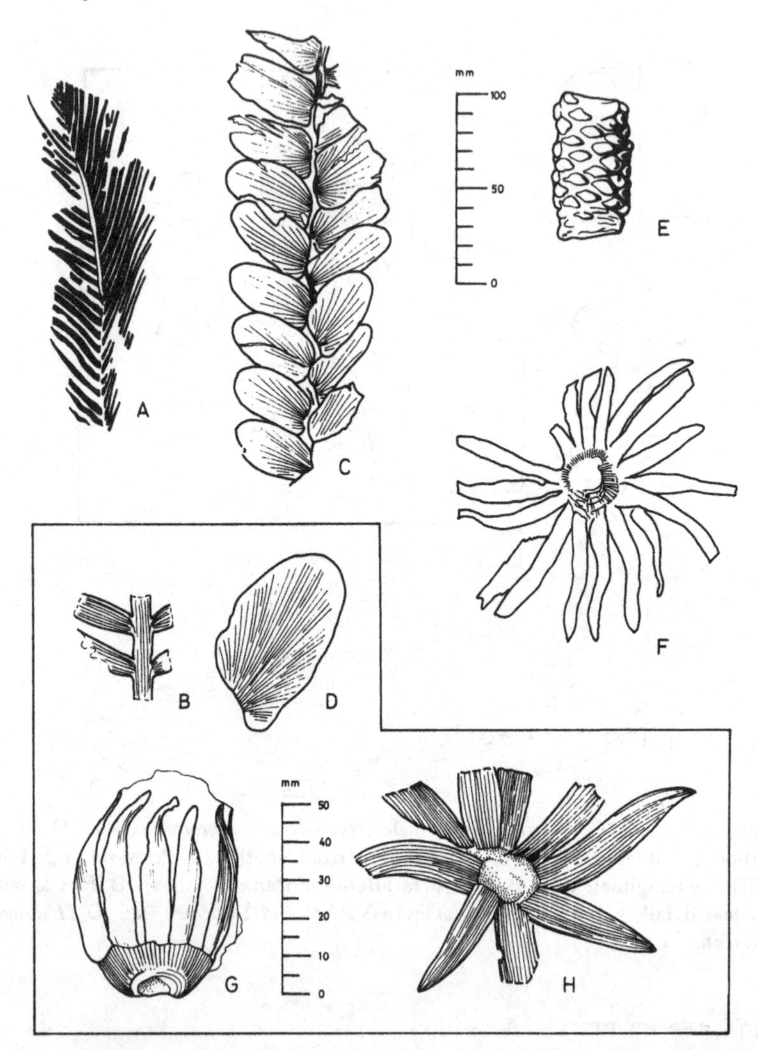

Figure 9.8. Early Cretaceous Benettitales of the group Williamsoniaceae. A–E, G, H from the Hastings Beds, Sussex, England, after Seward (1895); F from Primoria, Far East USSR, after Krychtofovich (1941). A *Zamiophyllum buchianus* (Ettingsh.), leaf ×0.25. B detail, ×0.5. C *Otozamites klipsteinii* (Dunker), leaf ×0.25. D detail, ×0.5. E *Bucklandia*, stem, ×0.25. F *Williamsonia pacifica* Krycht., flower, ×0.25. G *W. carruthersi*, unexpanded flower, ×0.5. H *W. carruthersi*, flower, ×0.5.

includes *Williamsonia carruthersi* from the English Wealden (fig. 9.8); of interest also is the well-preserved late Jurassic *W. scotica* from Scottish Kimmeridgian of Cromarty (Seward 1917). Wielandiellaceae are only represented by scattered records of *Nilssoniopteris*

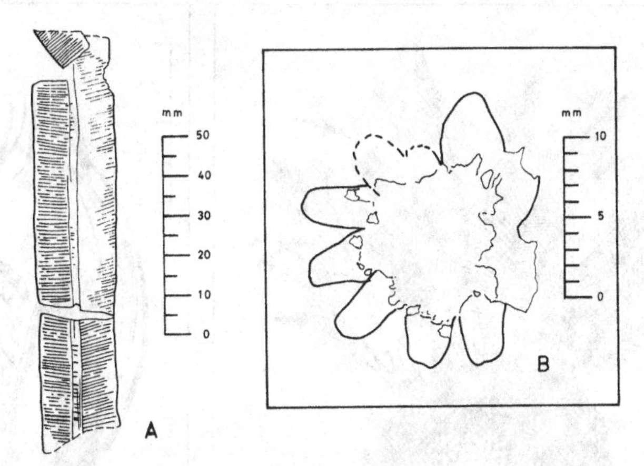

Figure 9.9. Early Cretaceous Benettitales of the group Wielandiellaceae. A *Nilssoniopteris rhitidorachis* (Krycht.), leaf from Ussuri suite, Far East USSR; Barremian age, after Krassilov (1967), ×0.5. B *Williamsoniella valdensis* Edwards 1921, from the Hastings Beds, Sussex, England; after Edwards (1921), ×2.

and by a single very small flower, *Williamsoniella valdensis* Edwards 1921 (fig. 9.9), of Berriasian age from Sussex, England. '*Cycadolepis*' from the English Wealden has been shown to have a Benettitalean cuticle (Oldham in press).

The Cycadeoidaceae are only recorded from rocks of late Jurassic and Cretaceous age. Their short petrified trunks (fig. 9.10) are well-known fossils of the early Cretaceous of Europe and particularly of North America; they are not yet known from the higher palaeolatitude of North-east Asia. Bisexual flowers are set sometimes amongst the persistent leaf-bases of the 'false' trunk (Wieland 1906, 1916) and are thus also petrified and have been studied in detail. Delevoryas (1968 and earlier) has suggested that some flowers were ovulate only, but has not figured any separate male flowers; that aspect of investigation is however difficult because of clear protandry which means preservation of very immature ovules in those flowers showing well-developed male organs. Wieland (1934) managed to show from the sectioning of a unique petrified crown of buds from the Aptian of the Isle of Wight that the leaves were stiff and pinnate, perhaps of the *Pseudocycas* type (fig. 9.10) which is common in early Cretaceous rocks as separate compressions. These compression leaves cannot be expected to, and do not, occur with the petrifactions; this however is a standard palaeobotanical difficulty. The well-drawn reconstructions by Wieland and by Delevoryas are

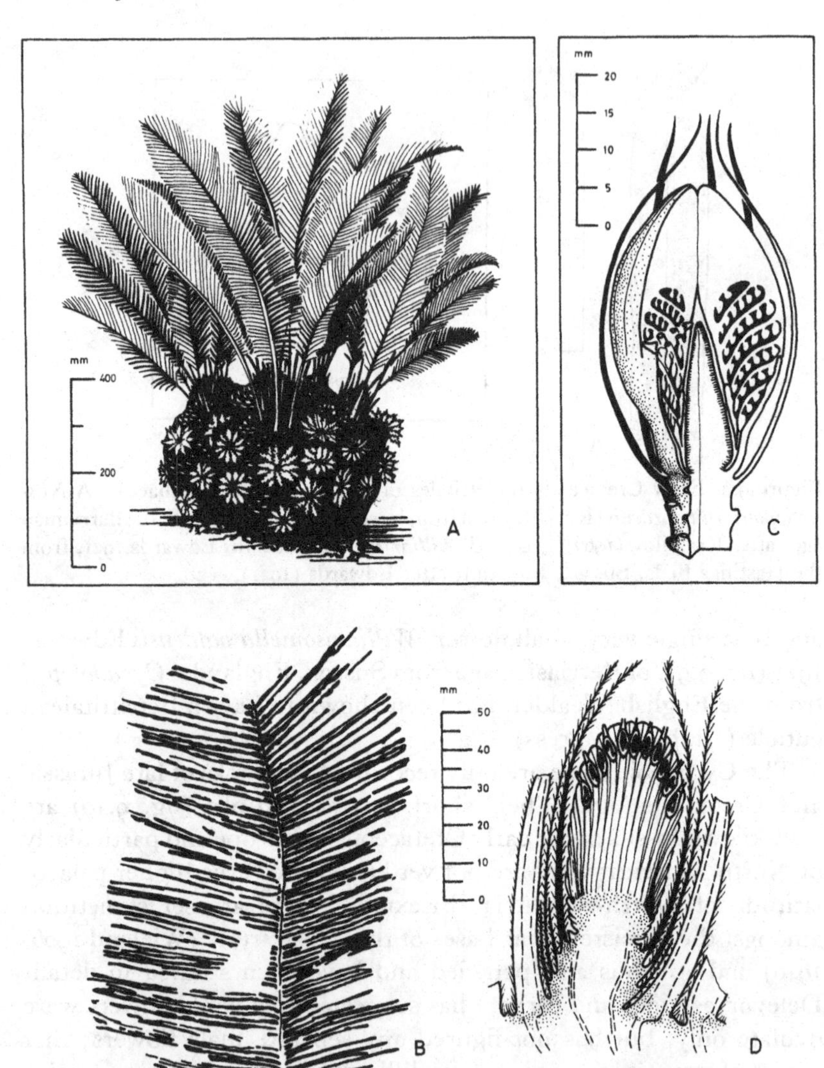

Figure 9.10. Early Cretaceous Benettitales of the group Cycadeoidaceae. A reconstruction of the plant *Cycadeoidea marshiana*; after Hirmer, ×1/16. B part of specimen of *Pseudocycas saportae* Seward, Sedgwick Museum, Cambridge (NH12); from Fairlight Clay, Lower Hastings Beds, Sussex; ×0.5. C reconstruction of protandrous bisexual flower; after Delevoryas (1968), ×0.5. D *Cycadeoidea Wielandii*; flower at ovule maturity, ×0.5.

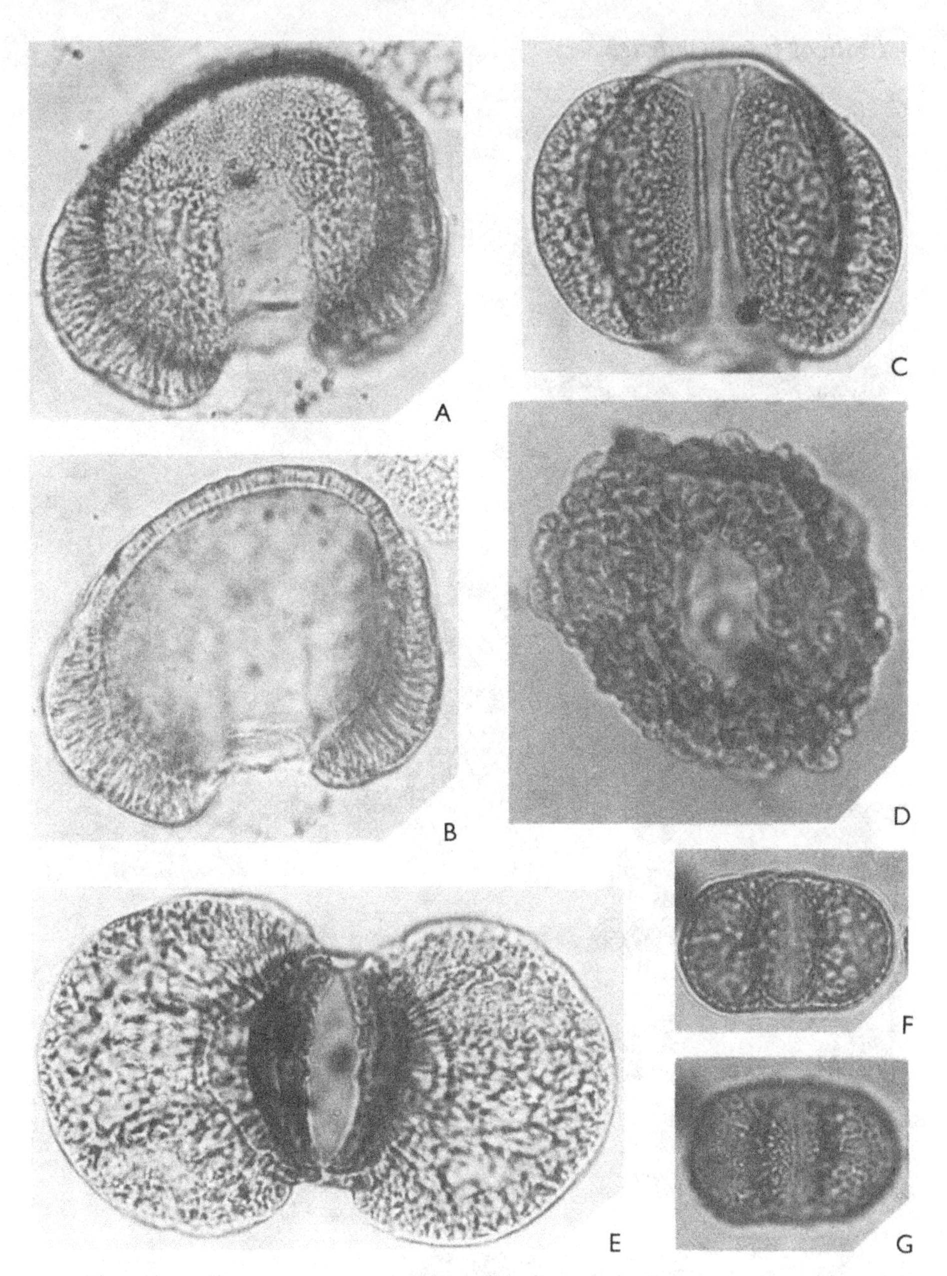

Figure 9.11. Saccate gymnospermous pollen from early Cretaceous assemblages; ×1000. A, B *Parvisaccites radiatus* Couper; depth 475 ft, Kingsclere borehole; late Barremian. A low focus. B mid focus. C *Abietinaepollenites* sp.; depth 1753½ ft, Warlingham borehole; Valanginian. D *Cerebripollenites mesozoicus*; depth 1740½ ft, Warlingham borehole; Valanginian. E *Podocarpidites* sp.; locality as for C, F, G *Vitreisporites pallidus* (Couper); locality NC1, Sandringham Sands, Norfolk, Berriasian. F mid focus. G high focus.

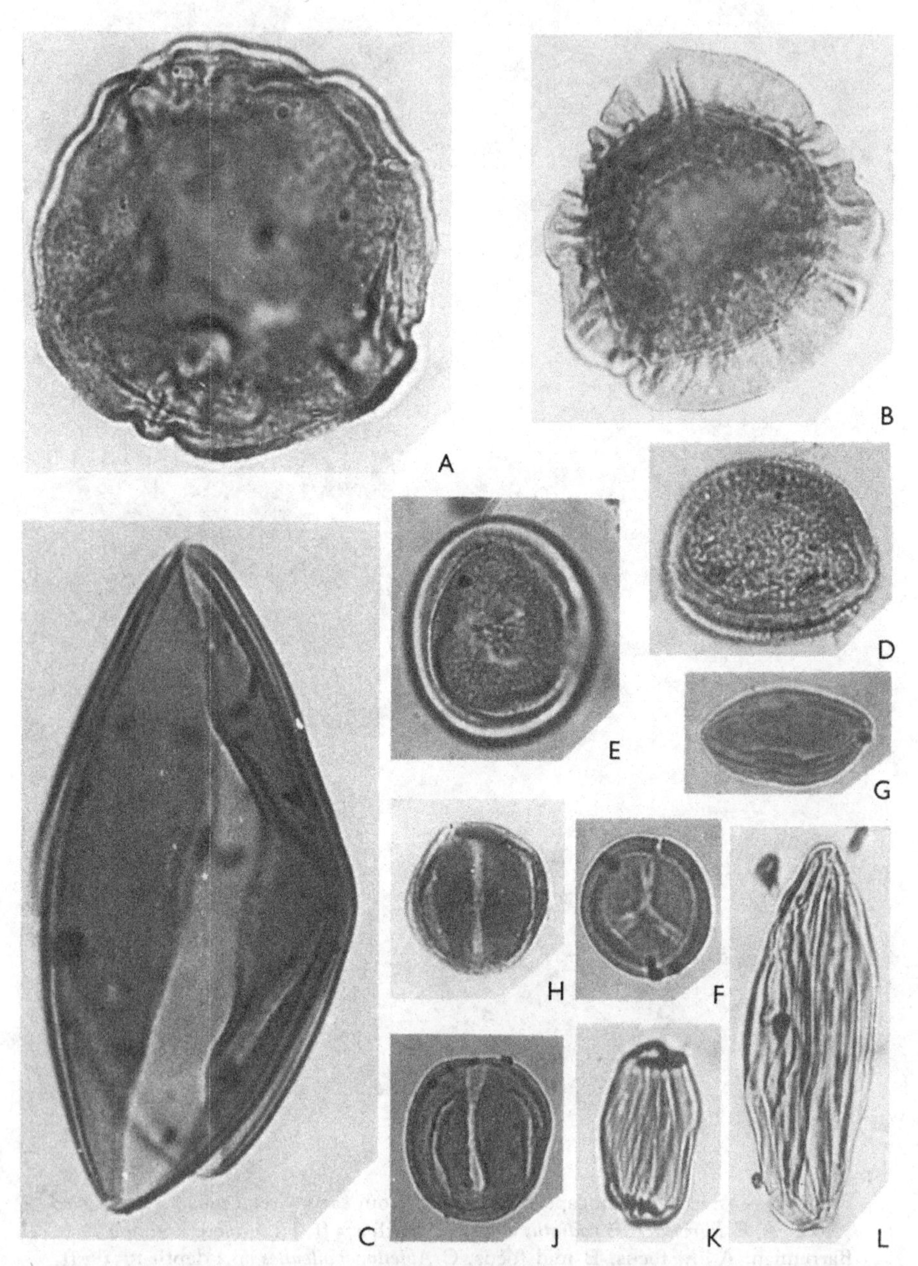

Figure 9.12. Gymnospermous pollen from early Cretaceous assemblages; ×1000. A–C from depth 1753½ ft, Warlingham borehole; Valanginian. A *Araucariacites* sp. B *Zonalapollenites dampieri* (Balme). C *Schizosporis* sp. D *Classopollis echinatus*; depth 1999 ft, Warlingham borehole, Berriasian. E, F *Classopollis* sp.; depth 931 ft,

Table 9.1. *Early Cretaceous gymnosperm pollen: roughly estimated abundance[a] and presumed provenance*

Dispersed pollen taxa	Estimated UK (30–40° KrN) abundance						Presumed megafossil link
	Berr.	Val.	Haut.	Barr.	Apt.	Alb.	
Monosulcites	1	1	1	1	1	1	'Ginkgoales' group; Nilssoniales; Benettitales
Eucommiidites	2	2	2	2	1	1	?
'*Ephedripites*'	0	0	0	1	2	2	Ephedraceae?
Araucariacites	2	2	2	2	1	1	Brachyphyll
Inaperturopollenites	2	3	3	2	2	2	?
Tsugaepollenites	2	2	2	2	1	1	?
Zonalapollenites	2	2	2	2	1	1	Linearphylls
Exesipollenites	3	3	3	2	2	2	?
Perinopollenites	1	1	1	1	1	1	Linearphylls
Spheripollenites	3	3	3	2	2	1	*Cupressinocladus*?
Classopollis	4	3	3	3	3	2	Brachyphylls
Abietinaepollenites	3	4	3	3	2	2	?
Parvisaccites	1	1	1	2	2	1	*Pinites* group?
Podocarpidites	1	1	1	1	1	1	—
Rugubivesiculites	0	0	0	0	1	2	?
Vitreisporites	2	2	2	2	1	1	*Caytonanthus*

[a] 0 = absent, 1 = present, 2 = common; 3 = abundant; 4 = universal.

widely used, but it is perhaps important to emphasise the difficulties with the information base. For instance, in addition to the uncertainty about the type of leaf, it is not clear whether these plants bore flowers once only, or more often; this and many other points can only be answered indirectly from material which must usually have been tumbled as logs in rivers before burial and petrifaction.

Legend to figure 12 (*contd*)
Winchester borehole; Hauterivian–Barremian. G *Monosulcites* sp.; depth 1740½ ft, Warlingham borehole; Valanginian. H, J *Eucommiidites minor*. H from depth 1843 ft, Warlingham borehole; Valanginian. J from depth 946 ft, Winchester borehole; Hauterivian–Barremian. K, L '*Ephedripites*' spp.; depth 1353½ ft, Warlingham borehole; Barremian.

9.11 DISPERSED MIOSPORES

Occurrences of pollen in palynologic assemblages (table 9.1) are perhaps best presented in three general categories, and illustrations of some common types are given in figs. 9.11 and 9.12.

Confirmation of the postulated plant distributions in Berriasian–Valanginian rocks of western Europe and eastern North America in a palaeolatitude of 30–40° KrN includes a predominance of *Classopollis* (Cheirolepidaceae and other Brachyphylls), *Abietinaepollenites* and other bisaccates (Linearphylls and more specifically designated conifers); there are frequent occurrences of *Spheripollenites, Perinopollenites, Zonalapollenites, Vitreisporites, Monosulcites* (which respectively may represent Cupressaceae, Taxodiaceae, brachyphylls, Caytoniales, cycadophytes and ginkgophytes). In Aptian–Albian time in the same areas *Tricolpites* and *Clavatipollenites* (probably early angiosperms), and *Ephedripites* (Chlamydosperms) count as confirmation.

Apparent anomalies include the relatively frequent *Araucariacites* occurrence without megafossil support; this may be due to false coupling of this form with living *Araucaria*, and it may be better explained as an as yet unspecified Mesozoic Brachyphyll. *Tsugaepollenites* is also relatively common, and the same kind of argument may apply. Both these forms are also common in Jurassic assemblages.

Of uncertain origin are *Exesipollenites* (sometimes abundant), *Inaperturopollenites* (also abundant) and *Eucommiidites* which occurs regularly. Totally unaccounted for are the pollen grains of Czekanowskiales, if they were indeed present in lower palaeolatitudes.

9.12 CONCLUSION

It is clear that for various reasons the critical details of early Cretaceous gymnosperms are little known because little attention has been paid to them. Future study of plant debris beds and of palynology will probably correct this in time, as long as evolutionary change is truly sought.

10 Early Cretaceous fossil evidence of angiosperm characters

Various fossils of early Cretaceous age show single characters or single organs of angiosperm type; the numbers of records increase steadily until in Turonian (late Cretaceous) time it appears to be certain that whole plants which would be acceptable as angiosperms were in existence as numerically important elements of the flora. The types of fossil evidence concerned and their relative importance have been discussed in chapter 3, and the traditional time-scale in chapter 7. The fossil pollen *Eucommiidites* was included under gymnosperms in chapters 8 and 9, and other pre-Cretaceous fossils suggesting angiosperm characters will be covered in chapter 13.

The many dispersed pieces of evidence recorded in this chapter are each discussed with reference to the simple succession statement presented in fig. 10.1; attention is drawn to any evidence in real or apparent conflict with this statement. In essence the critical time of the first statistically significant evidence of angiospermid characters is the Albian age/stage; everything earlier requires close examination for relevance. From Cenomanian time onwards the problems concern a satisfactory language for discussion of the integration of the many facts. For all these records it is necessary to assess time-correlations carefully, and particularly so outside western Europe, away from the rocks of the traditional time-scale.

10.1 EARLIEST CRETACEOUS EVIDENCE

From rocks of Berriasian, Valanginian and Hauterivian age there is very little record that is relevant. A single megafossil seed specimen has been described under the name *Carpolithus* from the marine Valanginian of Gigondas (Vaucluse, France) by Chandler (1958); unfortunately the specimen had been described in London before 1939 and was then returned to its French owner who could not subsequently be traced; the specimen must therefore be presumed lost and the observation cannot be considered significant until

Cretaceous stratigraphic scale (ages)	Flower	Fruit/seed	Pollen	Leaf	Wood
(later)		'Phytocrene'	Normapolles		
CENOMANIAN	'Magnolia' (petal)		+ +	+ +	
ALBIAN		*Kenella*	(tricolporoids) *Retimonocolpites* *Clavatipollenites* 2 *Tricolpites*	+	*Cantia*
APTIAN		*Onoana* 2		{ *Proteaephyllum* *Rogersia* *Ficophyllum* } *Acaciaephyllum*	*Aptiana*
(earlier)		*Nyssidium* *Onoana* 1	*Clavatipollenites* 1		{ *Woburnia* *Sabulia* } *Hythia*

Figure 10.1. Table showing the stratigraphic age of the earliest occurrences of fossils of isolated plant organs which have been accepted as evidence of angiosperm existence. +, ++ = occurrence of numerous fossils of these organs. *Onoana* species: 1, *O. californica*; 2, *O. nicanica*. *Clavatipollenites* species: 1, *C. hughesii*; 2, *C. rotundus*.

it is repeated. Palynologic studies of rocks of these ages have not revealed any consistent records of pollen with angiosperm characters. Millioud (1967) in a brief study obtained no angiosperm pollen from the traditional type localities of Valanginian and Hauterivian strata in the Jura mountains; the miospore assemblage was normal for the time, but as the strata were mainly marine this observation cannot perhaps be considered as conclusive evidence of absence. Examination of the English Wealden (Couper 1958, Hughes 1958, Hughes and Moody-Stuart 1967*a*, Batten 1973) and the German Wealden (Döring 1965) have revealed no angiosperm pollen of these ages. Burger (1966) recorded *Tricolporopollenites distinctus* Groot and Penny 1960 from some borehole sequences of Berriasian and Valanginian rocks of the eastern Netherlands. The time-correlation of the rocks both in general geologic and in palynologic terms appears to be satisfactory; the other miospores recorded fit perfectly well with normal Berriasian age occurrences in England, Germany and elsewhere. The grain of *T. distinctus* is small (diameter approximately 20 μm), and is described as 'indistinctly porate'; the number of specimens observed is not given but interpreting from the paper it may have been four, all from different rock samples. The holotype of the species concerned is of Cenomanian age and the species itself is only based on two specimens. The anomaly here is not so much the occurrence of angiosperm-type pollen but that it is reported as an advanced type not even found in the Albian early angiosperm occurrences; it is clearly necessary to repeat this observation independently and to obtain many more specimens, if necessary by processing much more material.

10.2 BARREMIAN TO ALBIAN FRUITS AND SEEDS

So far the most important discovery is the fruit *Onoana californica* described by Chandler and Axelrod (1961) from marine strata (Horsetown Group) of northern California (fig. 10.2*A*, *B*); these rocks are now believed (M. A. Murphy, personal communication 1970) to be of early Barremian age on the evidence of cephalopods, rather than of Hauterivian age as originally described. Unfortunately the locality is relatively remote and had not been re-collected; the single well-preserved calcite-impregnated specimen of a small unilocular fruit had a woody endocarp with some prominent pits which may have coincided with hair or spine bases. The new genus was attributed provisionally to the angiosperm family Icacinaceae which contains some well-preserved and fully described Eocene fossils (see chapter 11) but is not generally regarded as an early or primitive family on morphologic grounds; it is perhaps significant, however, that fruits of a type similar to *Onoana* from the Cenomanian–Turonian of Staten Island, New York, are among the very few adequate descriptions of Cretaceous fruits (Scott and Barghoorn 1958). The interest of the Californian discovery has been greatly increased by the description (Krassilov 1967) of several compression specimens of another smaller species, *O. nicanica*, from the Aptian strata of Sujfun, Southern Primorye, far-eastern USSR (Fig. 10.2*C*). Different and less fully described are compression material specimens of *Nyssidium* of Barremian age from the USSR (Samylina 1959, 1960, 1961); they represent an unknown strongly ribbed fruit with some resemblance to an angiosperm group, as suggested by the name.

Kenella is a small elongate bristle-covered fossil seed from the Albian of the Zyrianka river (Kolyma, north-eastern USSR); the description however of this compression material (Samylina 1968) is as yet minimal because the preservation is not good enough to provide cuticles (fig. 10.2).

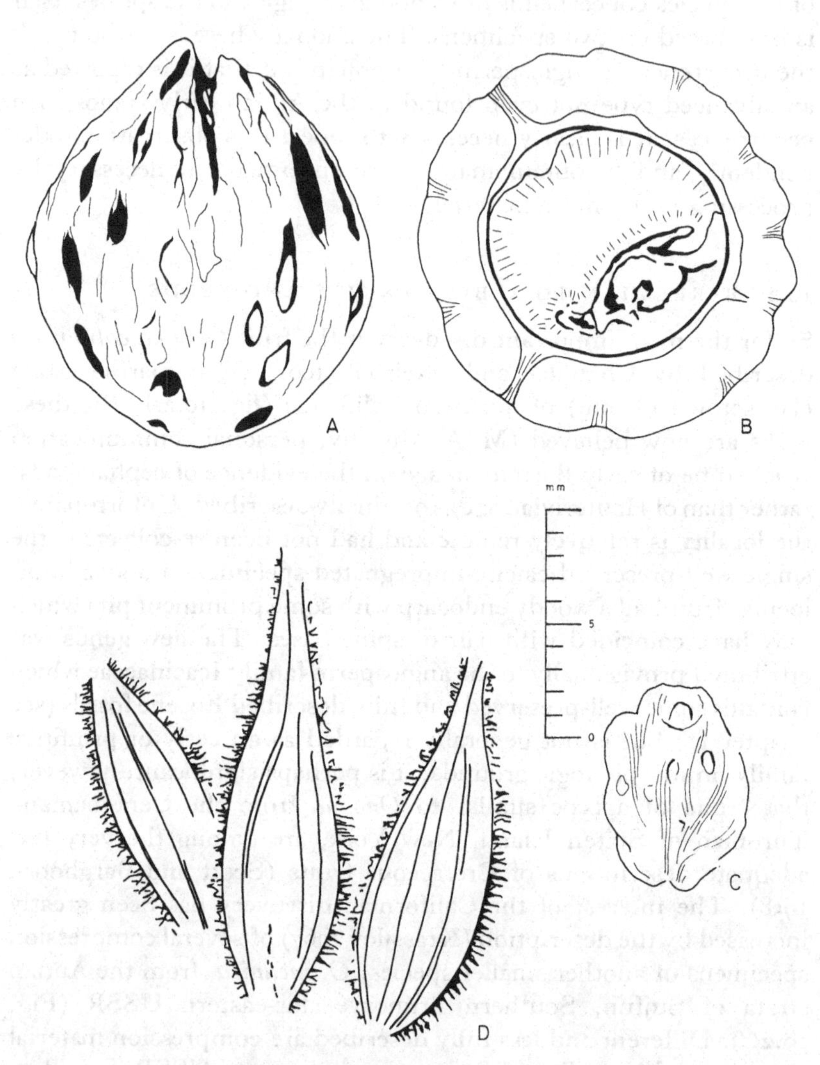

Figure 10.2. Cretaceous early angiosperm fruits. A *Onoana californica* Chandler and Axelrod 1961; from Horsetown Group, Northern California; age probably early Barremian (see text); after Chandler and Axelrod, ×3. B transverse section of same, ×3. C *Onoana nicanica* Krassilov 1967; compression material from Amur area, Far East USSR; age Aptian; after Krassilov, ×3. D *Kenella harrisiana* Samylina 1968; from Omsukchan, Kolyma River, East Siberia; age Albian; after Samylina, ×3.

The well-known megafossil floras of probable Albian and Cenomanian age (Blairmore, Patapsco, Portugal) are composed of leaves and include very few fruits and seeds. There are as yet no extensive Cretaceous petrifaction floras and Scott (1969) has commented on the lack of a sedimentation explanation for this. In fact if it were not for the records of *Onoana* there might almost be doubt as to whether angiospermy (of carpels) had been attained before mid to late Cenomanian times as represented by the Peruč Formation of Bohemia; from these rocks Velenovský and Viniklář (1926–1931) figured several enigmatic fruits in rich compression material now being restudied by Pacltová (1963) and others. On the other hand it seems probable that 'plant-fragment beds' have not yet been adequately searched in any relevant strata for such material.

10.3 POLLEN

The character of pollen clearly associated with the evolution of an angiosperm closed carpel and stigma is the development of numerous symmetrically arranged apertures as part of the provision for effective germination on the stigma; the other prominent general character, the elaboration of the outer part of the exine, was presumably naturally selected as a result of the various physical and biological problems of dispersal, transport, and stigma recognition.

In the following account first discussion is given to English and European descriptions simply because their time-correlation is better stated by current conventions than can be done for fossils from rocks elsewhere. The first well-documented occurrence is of *Clavatipollenites* (fig. 10.3) in which the actual monosulcate aperture is comparable with that of several gymnospermous plants but the outer exine (sexine) bears clavate processes closely arranged into a distinct 'tectum' covering the grain surface in an entirely new way. Kemp (1968) fully documented from the English late Barremian and Aptian strata what appears to be the earliest species (Couper 1958); this species, *C. hughesii*, is very small (length 12–25 μm) and the sculpture is 'retipilate' in that the heads of the clavae are not necessarily fused together. This species is also known from North and South American rocks of roughly similar age although in some of these cases the time-correlation is partly based on this fossil. Another slightly larger species (*C. rotundus* Kemp 1968) from the Albian of England and other correspondingly higher rocks elsewhere, is described as 'reticulate' (Doyle 1969) with the heads of the clavae fused to a reticulum. Such fossils bearing only one main angiospermous character may well be, as Couper (1958) thought, the kind of evolutionary link so often regarded as 'missing'. Various authors claim to have identified *Clavatipollenites* in earlier rocks: Pocock (1962) and Helal (1965) recorded such pollen from the late Jurassic but the specimens figured lack the critical exine features and can probably be referred to *Monosulcites* and *Eucommiidites* respectively. Schulz (1967) identified specimens from the early Jurassic of Poland which he attributed to the type species *C. hughesii*, but unfortunately this was done before Kemp's important redescription (1968) was available; there are clear differences in exine structure as well as in the shape and size, and it appears that some other grain was involved. All the other many occurrences are recorded from rocks that are probably of Barremian to Albian age, and there does not appear to be any genuine time-discrepancy between them even on account of latitude or palaeolatitude.

Asteropollis Hedlund and Norris 1968 is a Middle Albian grain with a similarly clavate exine to that of *Clavatipollenites* but is tetra- or pentachotomosulcate. There has been considerable interest in the possible evolutionary significance of grains with these apertures, as discussed by Chaloner (1970) for the trichotomocolpate grain. Doyle (1973) distinguished some species with a differentiation of fine and coarse reticulate areas of sculpture as 'monocotyledonoid monosulcates' which he included in *Retimonocolpites*, although his systematic study is not yet available. Neither an adequate number of specimens nor of localities has yet been studied to make this kind of speculation effective.

There has been some confusion of attribution of species between *Clavatipollenites* and *Liliacidites* Couper 1953, but Dettmann (1973) indicated a reasonable way of maintaining the generic distinction which will almost certainly be required as soon as records of such grains increase in number; it will be better to leave this position with some uncertainty for the present until such a distinction is more widely agreed on the basis of scan microscopy.

The first development of symmetrically arranged apertures (fig. 10.3) appears in the mid-Albian. The dated English example is *Tricolpites albiensis* of Kemp (1968); it is morphologically very close to *Tricolpopollenites micromunus* Groot and Penny 1960, and *Tricolpites minutus* Brenner 1963 from the Patapsco formation of Maryland, eastern USA, which are probably also Albian (but by palynologic correlation). There seems to be every good reason for retaining all

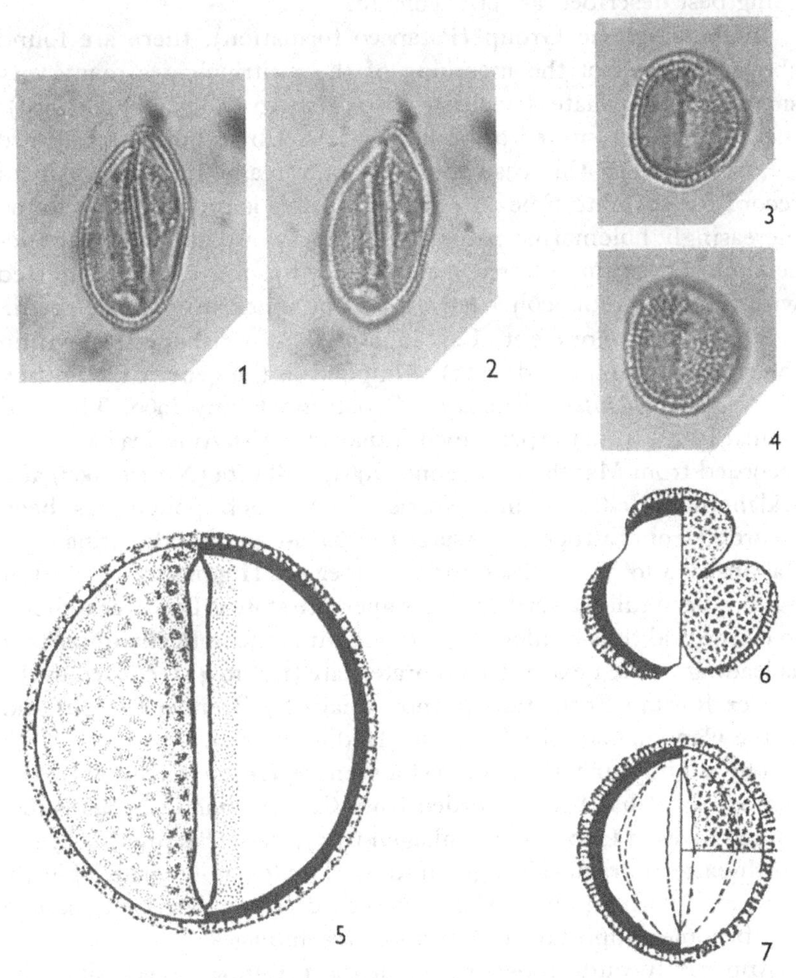

Figure 10.3. Presumed angiosperm pollen from Cretaceous Barremian rocks (depth 475 ft, Kingsclere borehole, southern England): *Clavatipollenites hughesii* Couper, ×1000. 1, distal view, low focus; 2, high focus. ×3, another grain, oblique view, mid focus; 4, high focus. (Rephotographed, after Kemp 1968.) Diagrams, ×2000 (after Kemp): 5, *C. rotundus* Kemp, distal face; left side showing reticulum in LO pattern; right side, optical section of exine; dark zone adjacent to colpus stippled in both. 6, 7, *Tricolpites albiensis* Kemp. 6, polar view; 7, equatorial view.

three names with their separate records; all three are prolate and very small (in the size range 10–20 μm) and are externally microreticulate from the arrangement of the clavate processes of the exine, being best described as retitricolpate.

In the Potomac Group (Patapsco formation), there are found above the level of the incoming of the retitricolpates many very small smooth (psilate) tricolpates (Brenner 1963, 1967*a*; Doyle 1969); the rocks are presumed to be in part late Albian but certainly also Cenomanian. In this case there is not yet any European Albian record for accurate time-correlation but this is probably due to the increasingly holomarine nature of the rocks in that part of the succession. These smooth tricolpates clearly to some extent co-existed with the early retitricolpates, but do not appear to precede them.

The first development of an equatorial germinal aperture within the colpus (tricolporoid state) is implied in the generic attribution of *Tricolpopollenites micromunus* Groot and Penny 1960. This and similar forms from the presumed Albian of North America have been recorded from Maryland (Brenner 1963), Alberta (Norris 1967) and Oklahoma (Hedlund and Norris 1968). Such pollen has been recorded from Europe and Asia as *Castanea*-type (Bolkhovitina 1959, Zaklinskaya 1963, Krutzsch 1957, Couper and Hughes 1963 etc.), but has not been fully described. This aperture state is however difficult to define and thus to identify, although it is theoretically important as leading to the next or tricolporate state (fig. 10.4) recorded in the Lower Raritan Formation (Cenomanian) by Brenner (1967*a*) and in the Cenomanian of Oklahoma (Hedlund 1966).

The slightly unexpected development of *Hexaporotricolpites* (tricolpodiorate) has been recorded from Cenomanian or perhaps just Albian rocks of Gabon (Boltenhagen 1963), Peru (Brenner 1968) and north-eastern Brazil (Herngreen 1974). The triporate form, logically correlated with an oblate shape, is recorded from late Cenomanian and becomes important in Turonian assemblages.

Apparently quite independent of the presumed succession discussed above, the pollen type representing the other logical correlation of form with germination on a stigma appeared in perhaps late Albian or certainly Cenomanian time. This was a spherical grain with numerous apertures regularly developed over the whole surface (fig. 10.4), such as *Multiporopollenites* of Jardiné and Magloire (1965), recorded from late Albian of West Africa. Doyle (1969) recorded rare polyporates from the Patapsco, and Brenner (1968) the same with tricolpates only from the Albian of Peru.

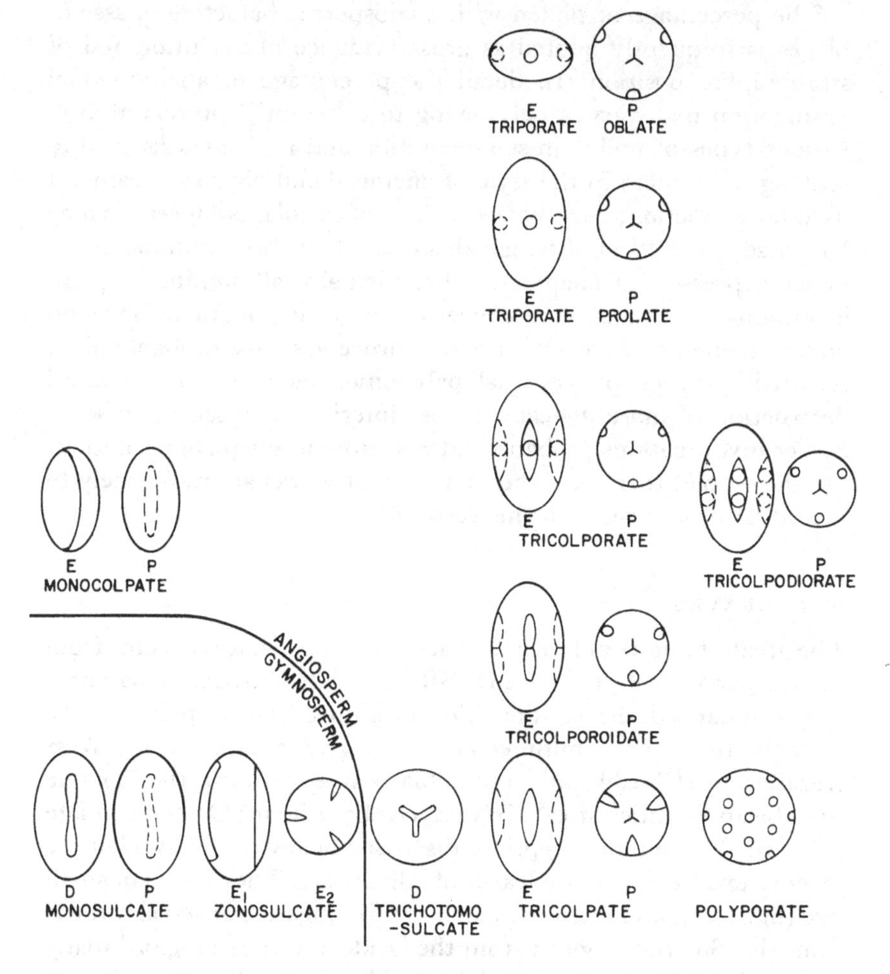

Figure 10.4. Diagrams of types of pollen found in rocks of mid-Cretaceous age; those placed above or to the right of the line are presumed to have been angiospermous, although the plants concerned may not have possessed 'angiospermid' characters in all organs. Monocolpate pollen (above the line) has tectate exine. Triporate forms are presumed to have evolved from tricolpate as suggested by their placing in the diagram, but the origin of other types is less certain. P, proximal aspect; D, distal aspect; E, equatorial aspect.

Pacltová (1966) recorded rare polyporates from the Cenomanian of Bohemia, and Hedlund (1966) from the Cenomanian of Oklahoma. Nothing is known of the tetrads of such grains and there is no indication of their origin although it was apparently during Albian time.

Critical fossil evidence

The percentage of pollen with angiosperm characters in assemblages is frequently quoted as gross evidence of evolution and of stratigraphic position. In detail the percentage in an individual preparation may vary widely owing to differential preservation of various types of pollen in sedimentation and in diagenesis, and to striking differences in the style of chemical and physical treatment used for extraction; basic palaeoecology of the plants concerned may have had some effect although this could have been minimal if the belief expressed in chapter 2, that virtually all significant plant evolution and fossils were concerned with aggradational lowland environments, is correct. Extraction processes have probably often resulted in the loss of very small palynomorphs, and in the chemical destruction of more delicate exines; inferior microscopes in some earlier investigations probably did not provide adequate resolution. On the whole, therefore, the more recent works are most likely to record accurate results in this respect.

10.4 LEAVES

The greatest number of early Cretaceous records of leaves come from six important localities in the USSR for which Samylina (1968, 1974) has summarised the results. The localities when replotted with palaeolatitudes (of Smith *et al.* 1973) give a range from West Kazakhstan (Chushkakul, mid-Albian) at 20° KrN to the Kolyma river basin localities at 53° KrN (Zyrianka, Albian; Omsuchan, late Albian). With minor exceptions discussed below, virtually all of the seventy taxa recorded are leaves of Albian age. They are all of small size (fig. 10.5), and many are narrow and elongate, particularly those from the Siberian region; from the 'Indo-European region' many are similar but a few are trilobate. Margins of leaves are either smooth (entire) or slightly denticulate. Most of the leaves differ in detail and have been placed in separate specific taxa; few happen to be preserved with cuticle. The generic names used in binominals mostly indicate supposed affinity with living families, and a few fossils have been inadvisedly placed in extant genera such as *Aralia*. Samylina (1968) expressed the opinion that they are all better placed in organ-genera.

Among the earliest floras of this age to be described (de Saporta 1894) were those of Portugal, subsequently revised by Teixeira (1948, 1950); of eleven floras throughout the Cretaceous, the relevant records (see fig. 10.6, 10.7) are: (*a*) Almargem, above Urgonian

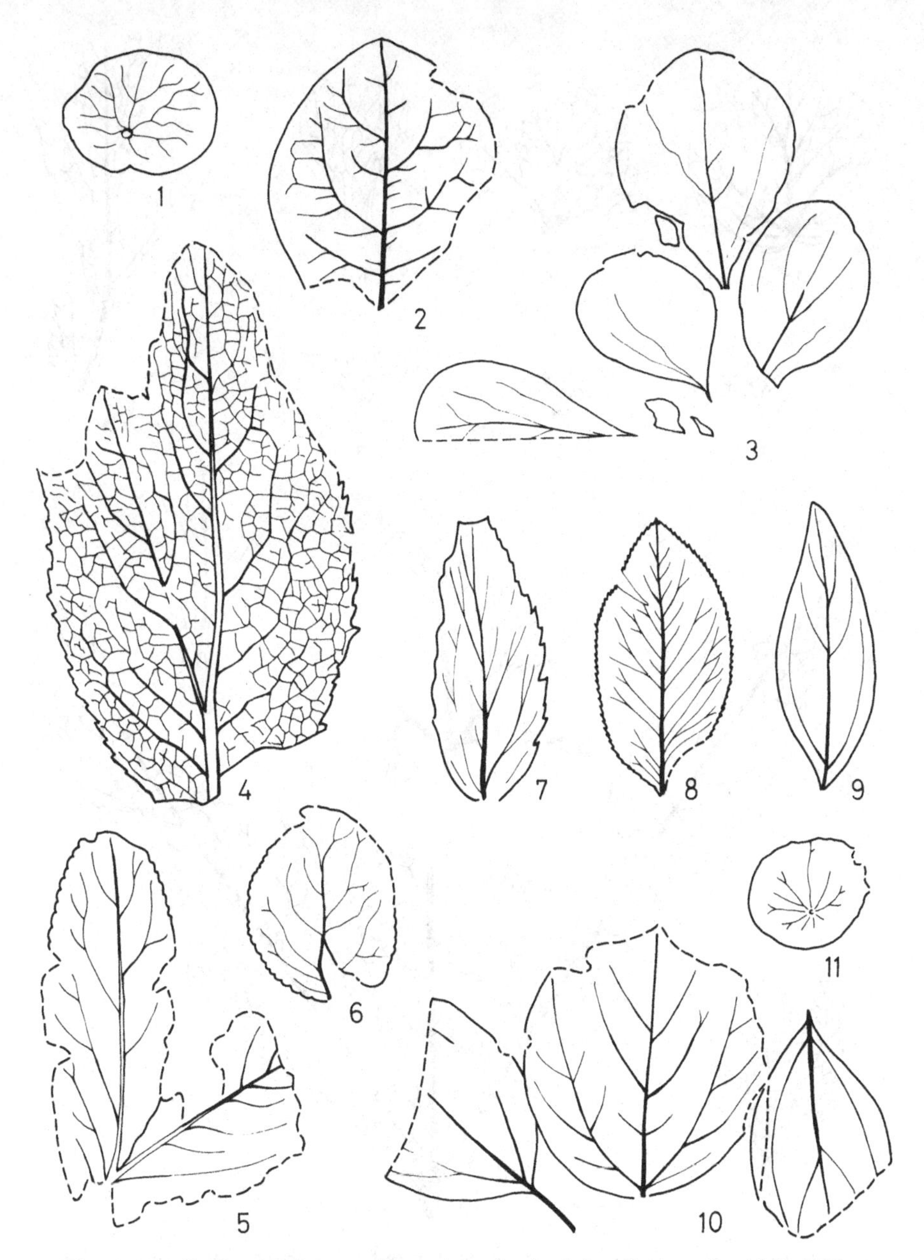

Figure 10.5. Earliest angiosperm leaves from Soviet Asia; all ×1. 1–6, middle Albian age, from Chushkakul, West Kazakhstan; after Vakhrameev (1952). 1, *Nelumbites minimus*. 2, *Ficus*(?) *tschuschkakulensis*. 3, *Leguminosites karatscheensis*. 4, *Celastrophyllum kazakhstanense*. 5, *Dicotylophyllum bilobatum*. 6, *Celastrophyllum ovale*. 7–11, Late Albian age, Toptan Suite, Omsukchan; after Samylina (1968). 7, *Celastrophyllum* aff. *hunteri* Ward. 8, *C. serrulatus*. 9, 10, *Cinnamomoides ievlevii*. 11, *Nelumbites minimus*.

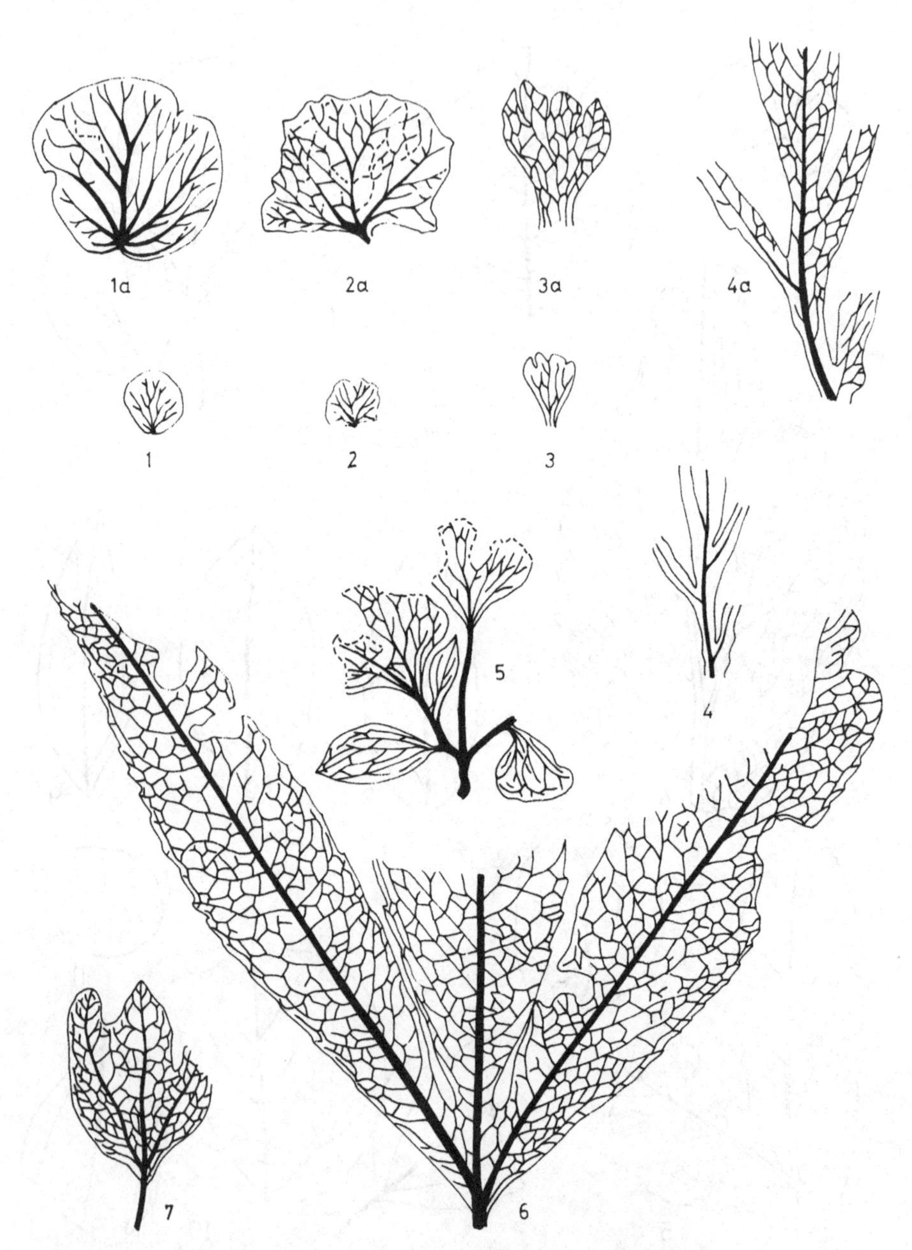

Figure 10.6. Cretaceous early angiosperm leaves from Portugal, after de Saporta (1894); ×1, except where otherwise indicated (1*a*, 2*a*, 3*a*, 4*a*). 1, 2, from Cercal; early Albian. 3–7, from Buarcos; Albian. 1, *Dicotylophyllum cerciforme* Sap. (*A*); 1*a*, ×3. 2, *D. hederaceum* Sap. (*A*); 2*a*, ×3. 3, *Proteophyllum leucospermoides* Sap. (*A*); 3*a*, ×2. 4, *P. dissectum* Sap. 4*a*, ×2. 5, *Adoxa praeatavia* Sap. (*A*). 6, *Aralia calamorpha* Sap. (*Ba*). 7, *Sassafras protophyllum* Sap. (*Ba*). For explanation of *A* and *Ba* see fig. 10.11.

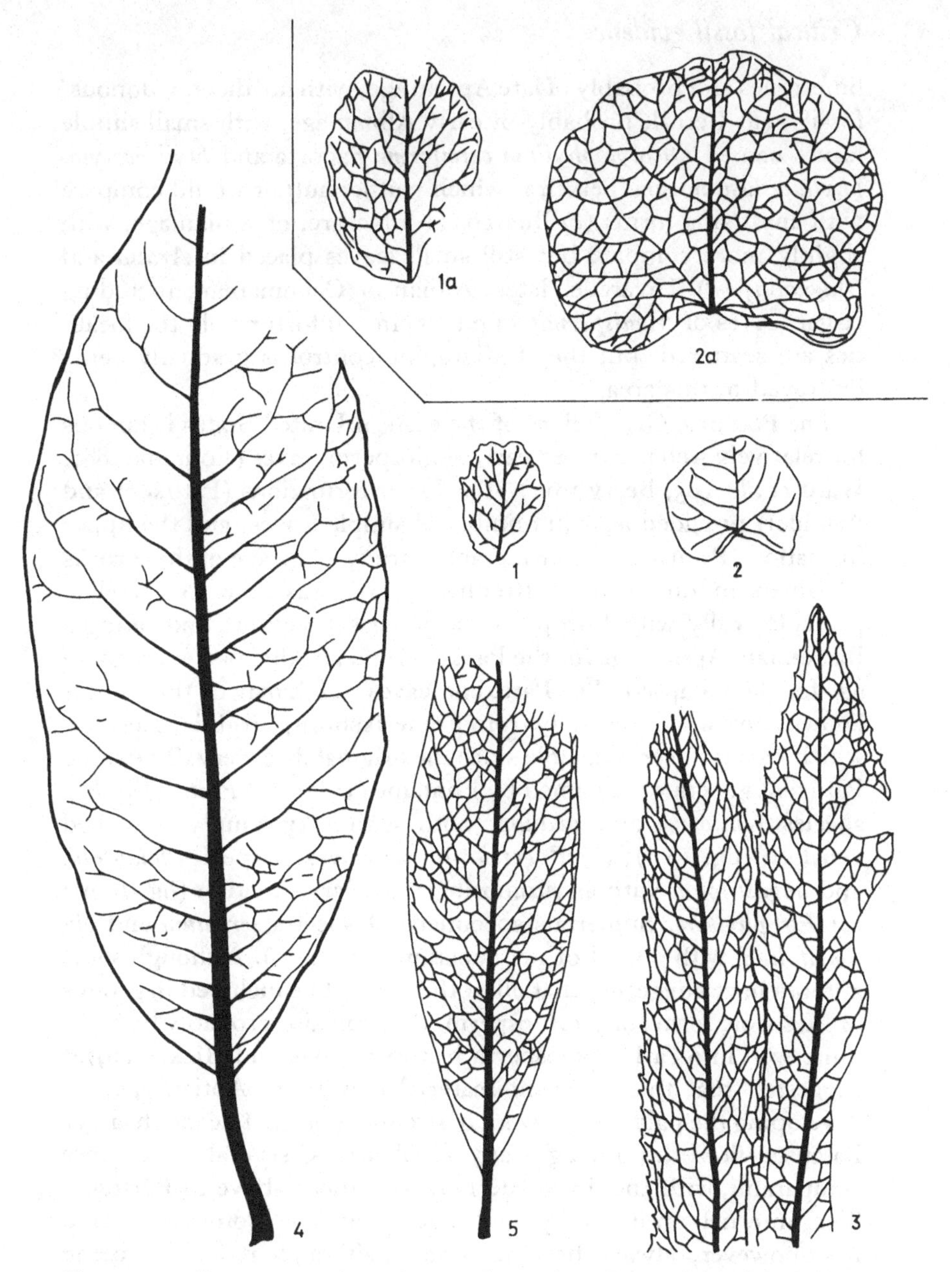

Figure 10.7. Cretaceous early angiosperm leaves from Portugal, after de Saporta (1894); ×1, except where otherwise indicated (1*a*, 2*a*). 1–3, from Buarcos; Albian. 1, *Myrsinophyllum revisendum* Sap. (*Bb*); 1*a*, ×2. 2, *Menispermites cercidifolius* Sap. (*C*); 2*a*, ×2. 3, *Salix infracretacea* Sap. (*Bb*). 4, *Magnolia palaeocretica* Sap. (*C*), from Bussaco; Cenomanian. 5, *Laurus palaeocretacica* Sap. (*C*), from Nazaré; late Albian. For explanation of *Bb*, *C* see fig. 10.11.

limestones and probably of late Aptian age, with no 'dicotyledonous' fossils; (*b*) Cercal, probably of early Albian age, with small simple leaves named *Dicotylophyllum cerciforme* Saporta and *Hydrocotylophyllum lusitanicum* Teixeira, which neither author would compare with any living form; (*c*) Buarcos and Nazaré, of Albian age, with slightly more complex but still small leaves placed in *Aralia* and *Magnolia*; (*d*) Tavarede, latest Albian or Cenomanian, including larger leaves of *Menispermites* up to 8 cm. Unfortunately the localities are scattered, but the stratigraphic control is gradually being improved in this area.

The Potomac Group flora of the eastern United States is famous for relatively well preserved early angiosperm leaves (Fontaine 1889, Ward *et al.* 1905, Berry 1911). The lower formations (Patuxent and Arundel) provided a small number of simple leaves, and the upper formation (Patapsco) a considerable variety. The age of these rocks remained in doubt until Brenner (1963) was able to correlate palynologically with European and other successions, indicating a Barremian–Aptian age for the Patuxent and an Albian–Cenomanian age for the Patapsco. The Patuxent leaves which may be the earliest known came almost entirely from Frederiksburg, Virginia, a locality which has not been available since the original discovery. The three Patuxent genera were named by Fontaine (1889): (*a*) *Proteaephyllum* species *reniforme* and *ovatum*, both with very simply branched venation (fig. 10.8); (*b*) *Rogersia*, linear leaves of species *longifolia* and *angustifolia*, with anastomosing secondary venation (fig. 10.9); (*c*) *Ficophyllum*, numerous specimens of species *serratum* and *oblongifolium* with a kind of net venation (fig. 10.10). Although some authors have suggested that all of these could be included in groups of Mesozoic forms of pteridosperm, they can also probably be fitted quite well into early angiosperm history; Doyle and Hickey (1972) suggested that this Patuxent material may be of Aptian age, by developing the palynologic work of Brenner (1963). The much larger Patapsco flora has among many small leaves, several larger ones comparable with the Tavarede flora and those above in Portugal. The internal stratigraphy and dating of the Potomac Group has, however, always been uncertain, although it is now being improved by palynologic study of borehole cores.

The Blairmore flora of the Albertan Rocky Mountains in the Crowsnest Pass Cretaceous coal-mining area was described by Bell (1956) who correlated the Lower Blairmore, with the single leaf-type *Sapindopsis angusta*, as Barremian–Aptian. It seems possible that

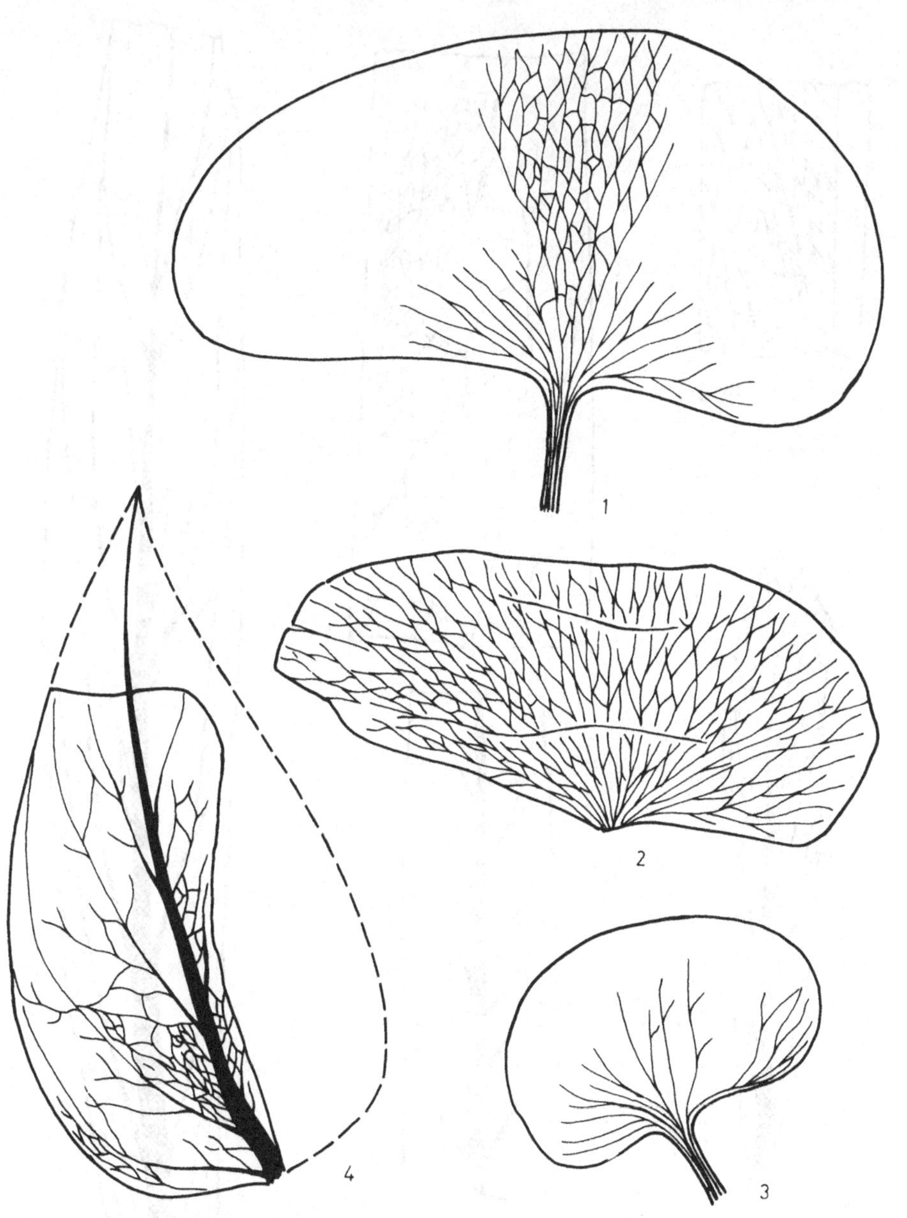

Figure 10.8. Cretaceous early angiosperm leaves of *Proteaephyllum* from eastern USA, after Fontaine (1889); locality Frederiksburg, Virginia; Patuxent Formation, Potomac Group; late Aptian–early Albian; ×1. 1–3, *Proteaephyllum reniforme* Font. (*A*). 4, *P. ovatum* Font. (*Bb*). For explanation of *A* and *Bb* see fig. 10.11.

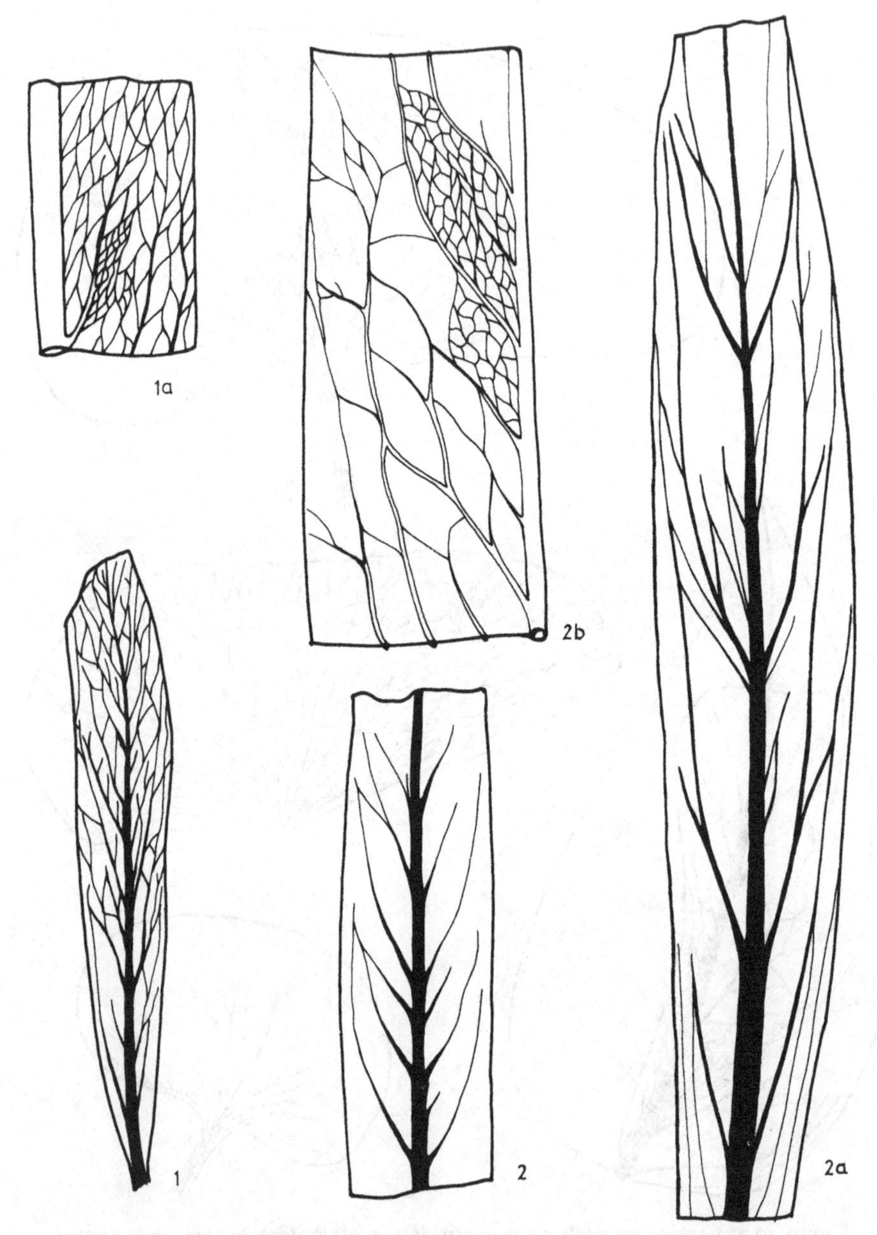

Figure 10.9. Cretaceous early angiosperm leaves of *Rogersia* from eastern USA, after Fontaine (1889); locality Frederiksburg, Virginia; Patuxent Formation, Potomac Group; late Aptian–early Albian; ×1 unless otherwise indicated (*1a*, *2a*, *2b*). 1, *Rogersia angustifolia* Font. (*Bb*); *1a*, ×4. 2, *R. longifolia* Font. (*Bb*); *2a*, ×2, *2b*, ×5. For explanation of *Bb* see fig. 10.11.

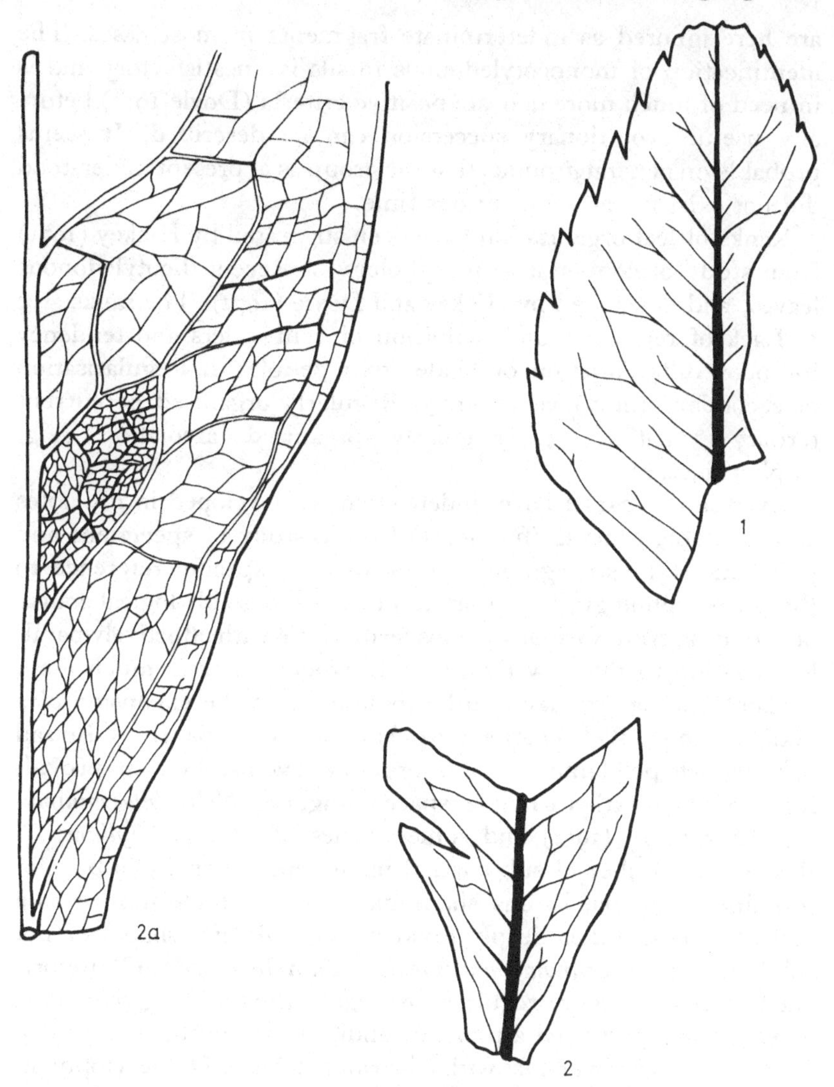

Figure 10.10. Cretaceous early angiosperm leaves of *Ficophyllum* from eastern USA, after Fontaine (1889); locality Fredericksburg, Virginia; Patuxent Formation, Potomac Group; late Aptian–early Albian; ×1, unless otherwise stated (*2a*). 1, *Ficophyllum serratum* Font. (*C*). 2, *F. oblongifolium* Font. (*Bb*); *2a*, ×3. For explanation of *Bb* and *C* see fig. 10.11.

this may be as late as Albian, and the Upper Blairmore which succeeds some barren strata has a flora with many taxa similar to those of the Patapsco and of perhaps Cenomanian age.

Monocotylodenous leaf records (de Saporta 1894 and many others) are here ignored as indeterminate fragments in most cases. The identification of monocotyledonous fossils is unsatisfactory and is in need of much more delicate positive criteria (Doyle 1973) before any useful evolutionary succession can be described. It seems probable on general grounds that the group as at present understood did not exist in early Cretaceous time.

Ranks of leaf organisation have been suggested by Hickey (1971) from study of comparative morphology of Recent dicotyledonous leaves, and developed by Hickey and Doyle (1972). The ranks are: 1. Lack of regularity and definition of vein orders and tendency for poor differentiation of blade from petiole; 2. Regularisation of secondary (only) venation; 3. Regularly organised percurrent tertiary venation; 4. Regularly organised areolation (e.g. Euphorbiaceae).

Over some years I have, independently, developed in my notes a comparable scheme (fig. 10.11) from a study of specimens and publications in stratigraphic order without specific reference to Recent morphology; it appears to present a possible logical course of evolution from various gymnosperm leaves with effectively parallel venation to the new light-weight expendable and much more efficient lamina. It may also be thought of as the expansion of a petiole with parallel venation into a lamina. The steps are: *A*. Round or wedge-shaped lamina without prominent veins, the veins providing a web-type rather irregular space filling, e.g. *Nelumbites minimus* of Vakhrameev (1952) and some species of *Dicotylophyllum*; *B*. Restricted number of sub-parallel main veins, with equal but not prominent laterals, lamina shape narrow as a single unit or as a palmate shape either deeply divided (e.g. '*Aralia*'-shape) or not subdivided (e.g. *Cinnamomoides ievlevii* Samylina 1968); *C*. Important lateral veins more regularly arranged, thus making possible a wider lamina and even a pinnate condition, including a variety of shapes, e.g. *Menispermites* with a broadened base; *D*. Development of a marginal vein; *E*. Full percurrent reticulation (in the sense of Dilcher 1974). The whole sequence can be viewed as degrees of efficiency in supply and drainage by venation, subject to the several obvious mechanical restraints. The shape may well have been evolved separately by a different selection for aerodynamic survival including petiole strength, evolution of the cuticle and stomata; and overall weight. It has been too readily assumed that a peltate shape indicated a floating aquatic plant and thus justifying the use of the name *Nelumbites* in several floras: it seems more likely to have been

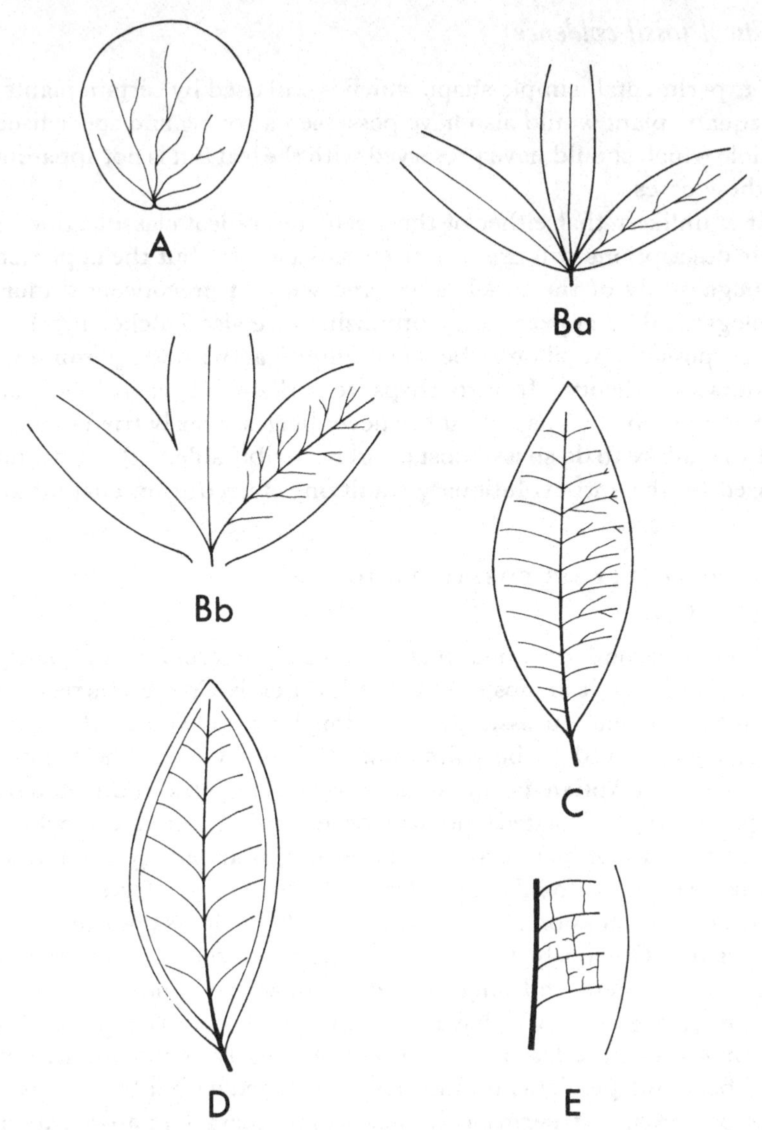

Figure 10.11. A possible series of early angiosperm leaf-morphology stages compiled by the author solely from study of published pre-Turonian leaf fossils; a complete sequence is not attempted and no botanical morphological theory has been used. *A* simple small rounded lamina with irregular ill-defined venation. *Ba* 'palmate' arrangement of three narrow laminae, each with a just distinguishable mid-vein and indistinct laterals. *Bb* arrangement as last but with fusion of the three laminae. *C* wide single lamina with strong mid-vein and relatively prominent although somewhat irregular laterals, giving a distinctly pinnate appearance. *D* wide lamina with marginal veins and more regular appearance. *E* 'reticulation' of minor veins developed. Not to scale although leaf size in general increases rapidly after Albian time.

an 'experimental' simple shape which is still used by certain plants; an aquatic plant would also have possessed a strong and specialised petiole which should have preserved with the leaf but is not apparent in these cases.

It is unlikely that either of these schemes of leaf classification or their descriptions will prove entirely satisfactory, but the approach through study of the fossils alone and without preconceived morphological ideas appears to be promising (see also Dilcher 1974); if at all possible it should be kept empirical with only minimal recourse to theory. It is perhaps remarkable, because such an approach is so obvious, that it has never been seriously tried before, but the failure to do so is almost entirely attributable to the restraint caused by the anti-evolutionary tradition referred to in chapter 2.

10.5 SCARCITY OF FOSSIL WOOD

10.5.1 *Wood*

Although secondary wood forms an easily preserved and fairly readily identifiable angiospermid fossil, the early Cretaceous records are few. The late Jurassic palms claimed by Tidwell *et al.* (1970, 1971) have proved to be palms but of Tertiary age. The famous description of Aptian dicotyledonous woods of southern England by Stopes (1915) is isolated and will be discussed separately below. There are some Albian records (Fliche 1905, Boureau 1954) but most angiosperm wood fossils are of late Cretaceous age or later.

After the discussion of the leaves given above this is not altogether surprising. Obviously the development of vessels to 'add to' an otherwise efficient and long-proved gymnosperm wood pattern of similar tracheids would have been in response to the successful exploitation of the leaf lamina. It could not have preceded it, nor have been independent; it must have developed with it from Albian time onwards, but perhaps the new vessel characters arose rather slowly.

10.5.2 *Lower Greensand woods*

These woods from four localities (Stopes 1915) consist of five separate specimens now in the Department of Palaeontology, British Museum (Natural History), which has made them available for study. Harris (1956) expressed doubts about their provenance; Hughes (1961*a*) suggested that use of such records should be suspended until at least one new confirmatory observation had been made. Casey (1961) from examination of sediment material of two of them considered the records entirely acceptable, with more confidence than was expressed by Stopes. The woods have been compared with somewhat advanced angiosperm types, and two of them are very well preserved. To explain the problems, the specimens will be taken by localities:

1. Luccombe Chine, Isle of Wight: coastal locality still available, late Aptian. Specimen V11517 *Aptiana radiata*, well preserved; no collection date or other information but rock matrix could agree with present outcrop. Small vessels 20–40 μm diameter, and multiseriate rays, could be considered primitive; strong growth rings. Gymnosperm wood abundant at locality but no other specimen of *Aptiana* yet found.

2. Igtham, north-west Kent: Folkestone Sands, early Albian. Specimen V13231 *Cantia arborescens*, well preserved; precise locality and collection date unknown, gift to museum collection sponsored by committee of Maidstone Museum 1913; water-worn, mollusc-bored; greensand matrix accepted (Casey 1961). Vessels 30–50 μm diameter, densely distributed; uniserial rays. Other fossils from area: gymnosperm woods and *Benettites*.

3. Maidstone, Kent; Hythe Beds, early Aptian. Specimen V13232 *Hythia elgari*, preservation fair; precise locality unknown, ex Maidstone Museum 1915. Isolated vessels 50–70 μm diameter, numerous multiseriate rays, wood contorted.

4. Woburn Sands, Bedfordshire: Lower Greensand, probably Aptian. Specimens B5452 *Woburnia porosa*, preservation fair and V5654 *Sabulia scottii*, preservation poor; no collection data or matrix. *Woburnia* very large 350 μm vessels, multiseriate rays, growth rings absent, supposed comparison with Dipterocarpaceae; *Sabulia* vessels 25–70 μm, uniseriate rays, many growth rings. Both these specimens were transferred unnumbered from the Botany to the Geology Department of the British Museum in 1898.

Both *Aptiana* and *Cantia* have locality doubts permanently expressed on their labels; the others have less precise localities. The palaeolatitude was approximately 35° KrN; the growth-ring information is inconsistent. It is just possible that the nature of the mollusc borings may prove to be distinctive and a guide to dating (see Yanin 1968). The urgent need is to find one further well-located specimen so that the matter may be profitably reopened; Luccombe Chine, Isle of Wight, appears to offer the most promise for further collecting.

10.6 FLOWERS

As with the corresponding leaves, angiosperm flowers are delicate and ephemeral compared with those of the Jurassic benettites. No flowers have yet been recorded from the early Cretaceous; the nearest is *Magnolia paleopetala* Hollick from the Dakota sandstone (probably Cenomanian), for which Leppik (1963) has enthusiastically reconstructed a complete flower from the single petal discovered.

10.7 CONCLUSION

The fullest evidence, which is palynologic, points to a Barremian or Albian beginning depending on whether *Clavatipollenites* is accepted; the fruit and seed evidence appears to indicate Barremian. The leaf evidence is predominantly Albian with a few isolated Aptian or even Barremian fossils; the wood supports this.

11 Late Cretaceous angiosperms

This chapter is intended to cover major changes and apparent anomalies but will not review in any detail the very large number of discovered fossils.

11.1 NOMENCLATURE OF FAMILIES

The great radiation of angiosperms from Cenomanian time onwards was originally documented in the period up to 1914 in the major leaf floras of Europe and North America. No new taxa above the genus were erected, and all the fossils were recorded on principle within about sixty extant families (53 Magnoliatae, 7 Liliatae); it was then subsequently noted by other writers that all these families had been in existence since mid-Cretaceous time, and that therefore the necessary evolution must have taken place before that time!

To a large extent this unsatisfactory state of affairs persists today; although modern revision has in many cases begun, the task is very long. The tentative schemes set out in section 10.4, or improvements on these, have not yet been used to any extent and no neutral family-level taxa have yet been erected.

Any review of the current situation must therefore use the existing taxa as has been done by Chesters *et al.* (1967), because although many of the attributions are irrelevant and misleading a language is necessary to discuss the fossils. Although it is only a refined version of the traditional process, a low (< 40) advancement index for families (Sporne 1969, 1972) can be used to select about a third of the sixty families which may be more nearly meaningful than the remainder (table 11.1). It may however be significant that only a very small number of the extant families cited as occurring in the Cretaceous are based on two or more different organs recorded from the same time interval; most relate to leaves studied long ago or occasionally to dispersed pollen.

Table 11.1. *List of those extant angiosperm families which are both mentioned as occurring in the Cretaceous (Chesters et al. 1967) and have an advancement index of less than 45 (Sporne 1972)*[a]

Below 30	Below 40	Below 45
Fagaceae 29	Anonaceae 32	Anacardiaceae 40
Magnoliaceae 27	Betulaceae 32	Casuarinaceae 44
Myrtaceae 27	Bombacaceae 35	Lauraceae 42
Platanaceae 29	Cercidiphyllaceae 35	Leguminosae 42
Rosaceae 27	Euphorbiaceae 30	Passifloraceae 40
	Hamamelidaceae 32	Rhamnaceae 42
	Juglandaceae 37	Salicaceae 41
	Menispermaceae 37	Sapotaceae 42
	Nymphaeaceae 34	Staphyleaceae 40
	Sapindaceae 35	Tiliaceae 40
	Sterculiaceae 32	Vitaceae 42
	Theaceae 32	

[a] The family Icacinaceae (see section 11.8) has an index 47; the family Celastraceae mentioned from the Aptian–Albian of the Potomac Group has an index 48.

11.2 EURASIAN CENOMANIAN FLORAS

Setting aside any difficulties due to lack of definition of stratigraphic scale divisions, there are facies difficulties which have prevented the dating of most of the well-known leaf floras by correlation with existing marine standards. This position is being steadily improved by the application of palynology. In no area however is the succession evidence complete and it is necessary to piece together data for megafossils from parts of Eurasia and North America and from different palaeolatitudes. Purely palynologic information will be dealt with subsequently.

In Europe the best-known flora is from the Cenomanian of Bohemia (near Prague; 25° KrN) with other records from the Cenomanian of Moravia and from Germany (Coniacian etc.), all on the southern border of the North European Chalk Sea; this sea margin was the only border for which the hinterland was tectonically active enough to provide suitable sedimentation. After the early work of the nineteenth century Velenovský and Viniklář (1926–31) carefully redescribed the Bohemian flora which is of mid to late Cenomanian age and unfortunately without good marine stratigraphic control below. Those authors recorded that the many pteridophytes and gymnosperms were accompanied by fructifications and were relatively easy to identify, whereas the angiosperm leaves, which were large (up to 15–20 cm), were unlike anything now living. Knobloch (1971) lists extensive new material with cuticles, collected since 1965 from new clay pits; about half appears to be simple angiosperm leaves with the principal types under the generic names *Araliphyllum*, *Debeya* (also perhaps araliform), '*Myrica*', *Myrtophyllum*, *Plataniphyllum* and *Magnoliaephyllum*, with no record of Liliatae (monocotyledons). Few of the observers agree on the affinity of any of these leaves with living families and in several cases, e.g. *Eucalyptus* sp., the implied attribution has been shown by cuticle studies to be erroneous (Pacltová 1961). In most cases also the dispersed pollen does not clearly support the leaf attributions. In the neutral descriptive terms used in section 10.4 above, most of the leaves fall into group *B* with some in *C*.

The floras of Portugal (also 25° KrN) on the Atlantic margin do not form a complete succession and have not been individually time-correlated with certainty. Tavarede provides a large orbicular *Menispermites* (group *C*) and is late Albian–Cenomanian; Bussaco (late Cenomanian) assemblage includes *Dewalquea* (now *Debeya*, see above), *Cinnamomum* and *Sassafras* (attributed to the Lauraceae which has an advancement index of 42), all (group *B*), and a *Magnoliaephyllum* species (*C*) now excluded from *Magnolia* on detail of venation.

Cenomanian leaf floras have been recorded from several localities throughout the USSR. Attributions to Lauraceae, Platanaceae (*Platanus* and *Credneria*) and Araliaceae are similar to those from Bohemia and Portugal. Vakhrameev *et al.* (1970, fig. 36) have drawn several province boundaries, but the actual localities as marked fall well enough into palaeolatitude belts. Krassilov (1973) has described a Coniacian flora with cuticles from Sakhalin (40° KrN) which includes *Platanus*, *Araliaephyllum*, *Debeya*, *Laurophyllum* and *Trochodendroides* (Hamamelidaceae); this is relatively similar to the previous flora except for the last genus, and for the serrated edges to some leaves, which character is normally regarded as a temperate zone feature. Krassilov suggests that all the fossils belong to extinct taxa except *Platanus*, and that the taxonomic and ecologic diversities were both low.

11.3 NORTH AMERICA: CENOMANIAN FLORAS

The Potomac–Raritan succession in Virginia, Maryland, New Jersey (palaeolatitude 40° KrN) is potentially the most complete for megafossils. The Patapsco (Upper Potomac) leaf flora (Berry 1911) has leaves of relatively small size which are subordinate in numbers to other plants. The principal taxa are '*Populus*' (attrib. Salicaceae), *Nelumbites* (attrib. Nymphaeales) (group *A*); *Sapindopsis variabilis* (Sapindaceae) abundant, *Celastrophyllum*, *Cissites* (Vitaceae), *Sassafras* (Lauraceae) and *Araliaephyllum* (all *B*). Only two rather dubious specimens of Liliatae (monocots.) are recorded. Because the only angiosperm pollen comprises several species of small thin prolate retipilate or reticulate tricolpates and one tricolporoidate, Brenner (1963) regarded the age as late Albian and this was accepted by Doyle (1969).

Unconformably over the Patapsco and after a time-gap lies the Raritan formation (see Doyle 1973) of which the Woodbridge Clay has been studied; in the pollen assemblages are tricolporates and triporates including *Complexiopollis* and *Atlantopollis* of the Normapolles group. This correlates with the Peruč of Bohemia which is mid and late Cenomanian (Pacltová 1966). It is not yet known how much is missing below in the early Cenomanian, and it is possible that only palynology of successions such as the French basin marginals will provide the necessary continuity (Laing 1975, in press).

Further west in the present United States is the Dakota group flora of Kansas described by Lesquereux (1892); in this case the Magnoliatae (dicotyledons) completely dominate the flora, many of the leaves are large (> 10 cm), and the estimated palaeolatitude is also about 40° KrN. Many of the leaves are lobate and plataniform in venation (*C*) despite a variety of generic names. Lower in the Kansas sequence is the Cheyenne Sandstone flora (Berry 1922) with the usual small-sized leaves. Other occurrences as far north as Blairmore in southern Alberta (palaeolatitude 65° KrN) have been claimed to correlate with those in Kansas. Bell (1956) believed the Upper Blairmore, although it contained large 'Platanus' (*C*) in an assemblage otherwise composed of *Araliaephyllum* and *Celastrophyllum* (*B*), correlated with the Cheyenne and was late Albian–Cenomanian in age. Doyle (1969) believed that the Upper Blairmore and Dakota of Kansas should be correlated, and that they may still be Albian (but also see May and Traverse 1973). This apparent confusion will probably be resolved as more material is restudied and more palynology co-ordinated, especially if the unusual Cretaceous palaeolatitudes are considered.

11.4 PALYNOLOGIC INVESTIGATION

The situation with palynologic work has been well summarised in different ways by Doyle (1969, 1973), Muller (1970) and Singh (1971). Muller, although he discussed the Cretaceous problem fairly, has built his presentation round 'the problems involved in identification of fossil pollen types with Recent taxa'. Doyle appeared specially keen to establish evolutionary trends in pollen grains, although he clearly appreciated that he was dealing with organs rather than organisms. Singh has brought all the North American information together with great care. Palynology of dispersed grains will have to be central to any ultimate history of early angiosperms because of the quantity of information potentially available, but it is doubtful whether understanding can progress much beyond the present level without a new neutral taxonomy for data-storage (see chaper 4), and more precise standards of recording.

Of present interest is timing by palaeolatitudes of the entry into the record of various distinct morphotypes of pollen, in an attempt to test for an equatorial origin and subsequent detectable poleward extension of range. As the various published records are arranged in Table 11.2, it seems possible that reti-tricolpates did originate first in the lowest palaeolatitudes, but not so clear that tricol*poroid*ates, polyporates or triporates followed them. It does also appear likely that tricolpo*dio*rates are of southern tropical origin at this time, to judge from the positions of Gabon and Peru. However, the interpretation is probably not significant in view of the state of data-recording and the stratigraphic correlation to as yet unstandardised reference scales. In addition the 10–20° KrN latitude belt plotted on the present geography appears to be unpromising for access and study.

11.5 CRETACEOUS MIOSPORES OF UNKNOWN AFFINITY

Although not by timing to be considered part of the main angiosperm development, the following cases at least suggest the existence of whole groups of plants as yet undiscovered.

The Aptian to Cenomanian ages produced some most unusual

Table 11.2. *First occurrences of the principal types of the earliest angiosperm pollen from localities arranged according to their Cretaceous palaeolatitudes*

Palaeo-latitude Kr	Location	Author	APTIAN	ALBIAN			CENOMANIAN	
				Early	Mid	Late	Early	Mid
65° N	Alberta	Singh 1971			Clavatipoll. Striate tricolpate Reti-tricolpate			
40° N	Oklahoma	Hedlund 1966 Hedlund and Norris 1968				Tricolpate Striate tricolpate Reti-tricolpate		Polyporate
40° N	Potomac	Brenner 1963 Doyle 1969, 1973	Clavatipollenites	Reti-tricolpate	Tricolporoid	Polyporate		Triporate
35° N	UK	Kemp 1968 Laing 1975*b*	Clavatipollenites			Reti-tricolpate	Tricolporate	Polyporate Triporate
25° N	Bohemia	Pacltová 1966				Trichotomosulcate		Polyporate
25° N	Portugal	Groot 1962						
5° N	Senegal	Jardiné and Magloire 1965		Reti-tricolpate Tricolporoid			Polyporate	Triporate
5° S	Côte d'Ivoire	Jardiné and Magloire 1965		Tricolporoid Reti-tricolpate			Polyporate	Triporate
10° S	N.E. Brazil	Müller 1966		Reti-tricolpate		Tricolporoid Polyporate		Triporate
15° S	N. Borneo	Muller 1968				Tricolpate Tricolporoid		
20° S	Peru	Brenner 1968				Tricolpate Polyporate		
25° S	Gabon	Belsky *et al.* 1965			Tricolpate Tricolpodiorate		Polyporate	Triporate
60° S	Australia	Detmann 1973				Tricolpate Clavatipollenites		Tricolporate

palynomorphs. Lammons (1970) described in impeccably neutral terms a sturdy stellate body formed on a spherical base from the Aptian of Peru. Stover (1963) and then many others (e.g. Jardiné 1967, Herngreen 1974) described a series of forms from mid-Albian (*Elaterosporites*), late Albian (*Galeacornea, Elaterocolpites*), to late Cenomanian (*Senegalsporites, Pemphixipollenites*) of the Cretaceous tropical area of northern Brazil, Gabon to Senegal, and Algeria; they possess long prominent processes interpreted as (incorporated) elaters, and in several cases no haptotypic features; they could represent a mixture of spores, pollen and plankton although the last appears unlikely; *Classopollis* only survives in the area until late Cenomanian.

The Normapolles group (Pflug 1953, Goczan *et al.* 1967) consists of oblate triporate pollen with complex and often protruding apertures. They appeared as simpler forms such as *Complexiopollis* in the mid-Cenomanian (Pacltová 1966), rapidly diversified in Turonian to Maestrichtian time, and disappeared by the Eocene. The occurrences are restricted to a province covering west Siberia, Europe and eastern North America, essentially between 20 and 40° KrN, with east and west limits approximately on lines of palaeolongitude (fig. 11.1); Doyle (1969) has suggested that these province boundaries were determined by the presence of epicontinental seas. The only published suggestion concerning affinity is that Normapolles were followed in Palaeogene time by Postnormapolles leading eventually to the 'Amentiferae'; this is morphologically feasible but does not explain the bizarre multi-layered apertures. It seems more profitable to seek explanation in the unique apertures themselves for which the function may have been exclusion of or control of entry of fluids; in the pteridophytes some perhaps comparable but simpler features occur in *Pyrobolospora, Arcellites, Perotrilites*, and *Balmeisporites*, which are all taken to have been organs of water plants. The idea for the Normapolles of pollenation over water may be worth investigating in view also of their provincial distribution; this is distinct from such water plants as *Nymphaea* with small simple pollen which are insect pollenated.

Complementary in northern hemisphere provincial distribution is the *Aquilapollenites, Wodehouseia* and *Proteacidites* assemblage in north-western North America, Japan and east Siberia. The time-range is from Coniacian to a rapid decline in the Palaeocene, but as with Normapolles the peak development is in the Maestrichtian. The occurrence of *Aquilapollenites* in Scotland (Martin 1968) is

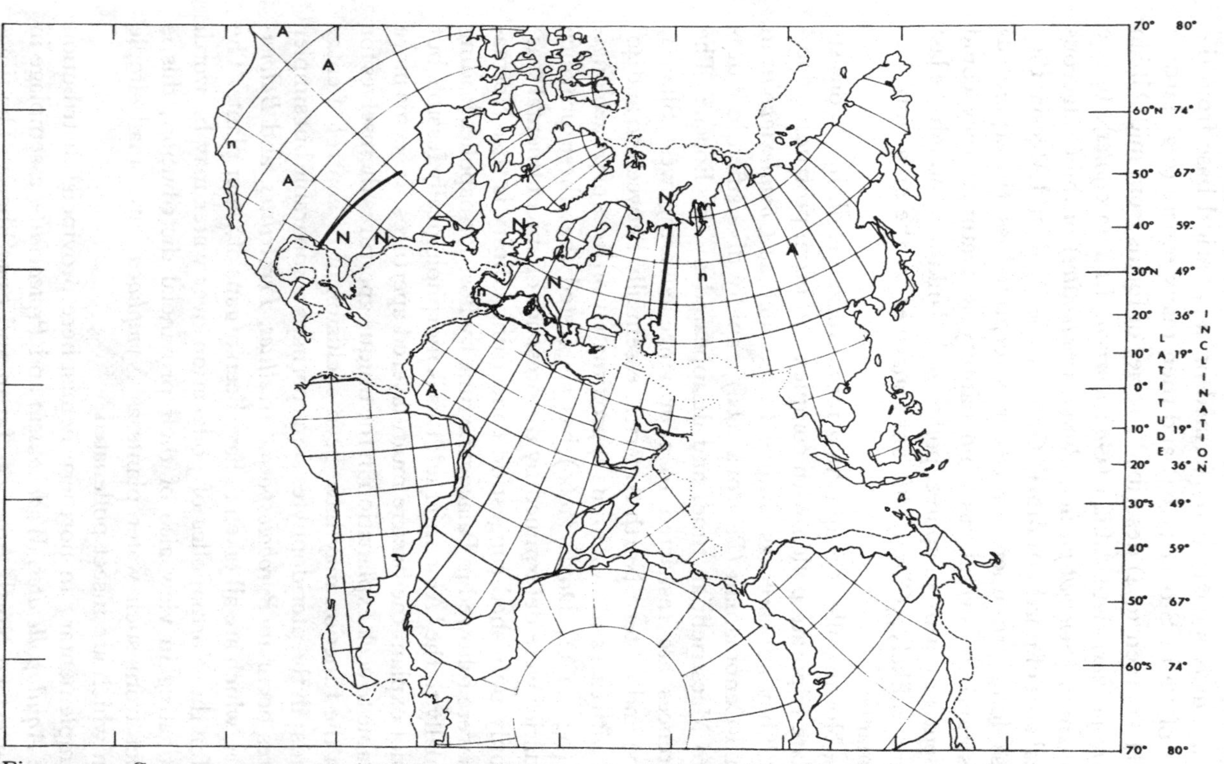

Figure 11.1. Cretaceous map with distribution of Normapolles (N) and of *Aquilapollenites* group (A) in the late Cretaceous northern palaeohemisphere. n, Normapolles of restricted variety. Information from Tschudy (1970) and earlier sources.

apparently anomalous but the province of distribution (fig. 11.1) is really north-polar and virtually all above 40° KrN (see Hughes 1973*a*); the grain itself has an irregular appearance due to several large processes bearing apertures, and its provenance is not known.

The Cretaceous southern hemisphere provinces of West Africa–South America, Australia–Antarctica–India, and Malesia (Borneo etc.) all carried district floras but not including any noticeably bizarre extinct elements.

11.6 WEST GREENLAND FLORAS

The work of Koch (1964) has made it clear that the Seward theory of polar origin of angiosperms, which was first based on Greenland data, was entirely misconceived because of stratigraphic errors. The pre-Cenomanian Kome flora contains no angiosperms; those reported by Heer (1874) were presumably 'float' from a late Cretaceous deposit above. The Upernivik Naes, Atane (?Coniacian) and Pautût (?Santonian) floras are all late Cretaceous. They are rich enough to be useful evidence at about 50° KrN, with the stratigraphy established.

11.7 LILIATAE (MONOCOTYLEDONS)

Because palm fossils have been claimed in both Triassic and Jurassic rocks, it is important to record the earliest megafossils, leaves and wood, which are from the Magothy flora in Maryland (Berry 1916) of Coniacian age; Muller (1970) reported *Nypa* pollen from the Maestrichtian, and other monocolpates, which may represent palms, a little earlier. Although the majority of palms at the present day are climbers or relatively small ground plants, the group has usually been considered in terms of standard solitary palms such as *Areca*. Doyle (1973) provided a more recent assessment of this problem.

All other Liliatae probably originated in Tertiary time as herbaceous successors. Graminae, which are of particular interest as originating and evolving with mammal herbivores, were apparently the first angiosperms to use silica on dry land and so to transform ground herbaceous vegetation in open non-forest high altitudes and latitudes; Recent articulate pteridophytes use silica but it is not clear when earlier members of the group adopted this habit.

11.8 FAMILY ICACINACEAE

This family is selected to illustrate some of the problems of attribution of fossils to Recent taxa because attributions to it include *Onoana* from Barremian to Albian strata, and also some late Cretaceous and Tertiary material.

Onoana californica was formally attributed by Chandler and Axelrod (1961) to '?Family Icacinaceae'. Miss Chandler, drawing on her great experience of Tertiary and Recent fruits, considered that 'it seemed necessary to establish a new genus'; she clearly expected such an early fruit to belong to 'the woody ranalean alliance' and was surprised to have to attribute it 'as representing a plant that had evolved well beyond that level'. The unilocular state, which had to be presumed, was also considered surprising.

The first fossil fruits attributed to the family were in the genus *Palaeophytocrene* from the Eocene London Clay (Reid and Chandler 1933) and from the Eocene Clarno Formation of Oregon (Scott 1954). *Icacinicarya* was erected for another fossil fruit from the London Clay, with a second species from the Palaeocene of Egypt (Chandler 1954). Meanwhile some entire leathery leaves from the Eocene of California were transferred from an old attribution to *Ficus* by Lesquereux, into the Recent genus *Phytocrene* by McGinitie (1941). Scott and Barghoorn (1958) then attributed some small endocarps from the Raritan (Cenomanian–Turonian) of Kreischersville, New York, also to a new fossil species of *Phytocrene*.

The living family Icacinaceae consists of about 400 species of tropical trees and lianes, with a 47% Sporne advancement index. Of about fifty genera, *Phytocrene* is one which happens to consist of about twenty lianes and shrubs in south-east Asia. The family is eurypalynous with tricolpates, triporates, polyporates and others. The living *Phytocrene* species appear to have pollen which Erdtman (1952) described for one species as 'oligo- and parvi-forate grains 16.5 μm diameter'; these might be termed very small polyporates (for other species see also Lobreau-Callen 1972). Attribution of fossil pollen to the Icacinaceae would thus be difficult and has not been attempted at least for the deposits mentioned above.

The wood characters for the living family are well documented (Metcalfe and Chalk 1950) but are diverse and have not, apparently, been recognised fossil. *Phytocrene* wood is particularly specialised to the liane habit.

The family Icacinaceae has itself been placed in various orders of the subclass Rosidae such as Santalales, Celastrales, or Sapindales, and is regarded as a relatively advanced group.

What has been done has been done with wisdom at the time and in good faith. It seems necessary, however, to decide whether the cumulative effect of what is recorded above amounts to an investigation of evolution; if not, what arrangements should be made for data-handling of evidence for the many hundreds of other angiosperm families?

11.9 THE FOSSILS

The late Cretaceous angiosperm megafossils above amount without the pollen to a very large number of records. Most of these need revision but in something more than the sense of attribution to living families so that they may be named. Wood has been relatively neglected but this is fortunate as traditional handling of it would have produced even more confusion than in leaves and pollen.

In gross numbers present in assemblages angiosperm leaves usually predominate over all others by Turonian time, but the percentage increase per age from Albian through Maestrichtian is on the whole steady. This forms part of the argument for steadiness in selection factors acting on the plants.

As a suggestion for the Cretaceous, any new revision should probably include the abandonment of all names of extant taxa below the 'order', and the building of an entirely morphographic, stratigraphic and geographic data base without further interpretations (see chapter 16). The passive weight of botanical custom behind the handling of all aspects of all this late Cretaceous material has been the main cause of continuing the myth of an 'abominable mystery'.

PART 4

Conclusion from evidence

12 Theory of angiosperm origin and early evolution

12.1 THE CURRENT FAILURE OF THEORY

As suggested in chapters 1 and 2 one of the main reasons for the great difficulty experienced over the problem of the origin of angiosperms lies in the general method of approach which has amounted to an unconscious denial of evolution, while at the same time a somewhat indiscriminate searching for detailed evidence of evolution continued. All present living angiosperms must have evolved from some Eocene angiosperms which were different, and they in turn from Maestrichtian and further back from Cenomanian plants that were different again. However, almost every author who has referred to this problem mentions only modern angiosperm families that 'were established by mid-Cretaceous time'. Virtually no supra-generic taxa have yet been erected for fossil angiosperms that do not extend through to (or rather back from) the present day. Even the several Cretaceous and the relatively few Tertiary palynologists who wisely use taxa and names based on purely morphographic features, appear to regard their action as temporary until a suitable living family can be found to accommodate their fossils. The often quoted 'failure' of the known fossil evidence to clarify the problem against this background has led to the conclusion that the fossils are totally inadequate and that classification of all angiosperms should proceed as before (plus modern technology) without wasting time on fossils (e.g. Davis and Heywood 1966).

The solution here offered, therefore, is to excise as completely as possible all traces of this methodological defeat, and to concentrate all effort on interpreting the abundant fossil material as it is; and in addition to devise a data-handling code that is sufficiently sound to make possible some computer manipulation for a solution that is likely in the end to be beyond the scope of the small number of diligent palaeobotanists (and of their files) who are likely to be engaged on this task.

12.2 THE FOSSIL RECORD

The normal supply of most land-plant fossils was from vegetation that grew close to sea level on aggradational land such as deltas. The most intense natural selection by biologic competition took place in these same favoured areas. The fossil record may therefore be expected to give a reasonably accurate history of the main course of land-plant evolution.

Leaves and wood, and pollen and fruits, make up the great majority of available relevant fossils. Evolution and classification must therefore be worked out by using characters derived from these organs alone. Fossil pollen occurs in such quantities that problems arise with selection of the observations to be made and with their significance; the theory involved in such selection must apply equally to all fossils even if some of them may at times be scarce in the rocks. Dispersed miospore palynology, although it deals only with one type of plant organ, has become the guide and the lead into the whole investigation. Palynology is only likely to be superseded in this role by a set of features providing a greater total number of character observations; although it is by no means available yet this could arise in the field of palaeo-organic chemistry.

12.3 ANGIOSPERM DEFINITION

It has often been suggested before that the various plant organs should not have been expected to develop simultaneously all the characters known now as angiospermid. Further, it is important to remember that the *accepted* definitions of angiospermy are all based on the sum of observations made only on a large amount of Recent material. A Cenomanian definition of angiospermy might well have been strikingly different in the direction of simplicity, because it could only have been based on perhaps two or three families with *A*-, *B*- and *C*-type leaf venation (see section 10.4), and with simple monosulcate and tricolpate pollen. The group might not even have been termed 'angiosperm', but perhaps by some such name as the 'lamina-stigmatics'; the group might not have been considered homogeneous nor particularly distinct from the Benettitales, the Caytoniales or the Czekanowskiales which still existed in Cenomanian time. These last groups, although extinct, have now all been loosely called 'gymnosperms' by making use of diagnostic characters largely based on the relatively small number (640 species) of living plant survivors; how could these Cretaceous plants have been classified using only those characters that were available up to Albian time?

This approach will not be further developed here because it is too destructive of the only available language of communication, but it appears to have promise.

12.4 ANGIOSPERMID CHARACTERS IN FOSSILS

The Albian 'angiosperm' leaf consisted of little more than the production from a petiole of a lamina by webbing between relatively irregular (but not obviously dichotomous) main veins. The leaf shapes were initially rounded but later simply narrow and lanceolate, but all the leaves were small; the choice of shape was presumably limited, as it is now, by aerodynamic circumstances in open areas. Dense canopy forest almost certainly had not yet appeared in Mesozoic gymnosperm-dominated vegetation. Some leaf margins were serrated, this presumably being a selection (as now) against accidental tearing in more open sites. The fact that many of these shapes resemble those of living plant leaves is to be expected when the range of physical possibilities for a leaf lamina at any time is limited. What is not similar, however, is the venation which is very poorly displayed in many Albian fossils; this is not entirely due to indifferent preservation and in many cases is due to a truly primitive state of irregular development. This all appears to be normal and as might be expected in the early stages of evolution of a complex lamina.

A new development of woody conducting tissue was probably essential as soon as the lamina had proved to be successful. The information on evolution of wood is still fragmentary especially if the English Lower Greensand woods are ignored, as they probably should be for the present. A theoretical morphologic series can easily be imagined, but it will be necessary to wait for discovery of the facts from fossils. Perhaps they will be found more easily if there are no 'well-established living families' in mind.

The state of development of ovule, seed and fruit is very hard to determine from the fossils at present known. *Onoana*, *Nyssidium* and *Kenella* present only a little more than the external appearance of the fruit. There is no trace of structures such as those seen in the living 'primitives' *Drimys* and *Degeneria*, which suggest an ideal theoretical method of carpel formation; as the few known fossils bear no resemblance to these genera and if anything suggest more normal structures, perhaps it is best to omit these 'primitive' genera from consideration. There is as yet no indication that they are anything but quite recently evolved, despite their interesting morphology.

The nature of the earliest angiosperm pollen is rapidly becoming well documented in numerous palynologic assemblage successions which will eventually be satisfactorily time-correlated. Although a monosulcate grain such as *Clavatipollenites* does not itself indicate much advance on gymnospermy, if any, *Tricolpites* of the Albian onwards would apparently indicate the existence of a true stigma as a 'purpose' for the production of a symmetrically aperturate pollen. Whether a tectate exine performed a physiological or physical set of functions, or both, is not yet known but it certainly suggests important evolutionary developments in mid-Cretaceous time; whether cycadophyte grains or *Classopollis* had paralleled some of the details earlier is not important.

Some petrifaction preservations are urgently needed to clarify the ovule-to-fruit problem. They should be sought in areas of Cretaceous sediment with interbedded pyroclastics.

12.5 MESOZOIC LAND-PLANT EVOLUTION

In Triassic and Jurassic time there were certainly no angiosperms in any sense now recognised; no angiospermid characters have been substantiated (see chapter 13). The very diverse existing seed-plants of those periods, now loosely called 'gymnosperms', had already diverged widely. Quite independently, at least the Caytoniales and *Leptostrobus* (Czekanowskiales) developed some grouping of ovules under one cover, a step which was as important as the subsequent one of providing a closure and a stigmatic surface (together); the advantage was presumably in distribution, perhaps through the agency of small vertebrates which might have taken an edible fruit and thus dispersed small seeds. The Caytoniales also developed a limited leaf lamina on a pattern of anastomosing venation (*Sagenopteris*) that had been used before (e.g. Carboniferous *Lonchopteris*, Permian *Glossopteris*) without leading to any obvious further radiation. In the Benettitales of the late Triassic more definite integration with animal life was correlated with a visually attractant bisexual flower, e.g. *Sturianthus* from Lunz; the grouping of ovules and inter-seminal scales amounted in this case to little more than a condensed cone with large numbers of small seeds leading to accidental distribution by animals; pollenation of these many ovules in a relatively small space was presumably effected crudely by wandering pollen-eating insects attracted visually or even additionally by odour; crossing could have been provided for by the protandry which is suspected, and the pollenation and distribution visitors could have been different. Two independent pollen types each had some kind of effectively universal aperture. *Classopollis* may be regarded as possessing solely a distal pore, which is not different in effect from a monosulcate condition, but there is in addition the distal rimula which may have been a zonosulcus; there is a suspicion that the pollen may have been functional in tetrads; the plant is otherwise known to have been a normal Jurassic conifer. *Eucommiidites* is monosulcate and apparently also zonosulcate, providing in effect a close approach to a tricolpate arrangement; the function of this feature, however, is not at all clear because the grains have on several separate occasions been found in micropyle and pollen chamber of a gymnospermous ovule, and on one occasion housed in a small cone.

All these developments are in different plant groups, all of which are extinct. The pollen developments are unexplained.

In very early Cretaceous (Berriasian to Hauterivian) time there were also no angiosperms. All the specialised (gymnospermous) seed-plant groups mentioned in the last paragraph are again recorded but of these the Czekanowskiales appear to have radiated to a north high-latitude distribution by this time. The Caytoniales appear to have continued universally but known Cretaceous details only include new leaf species with minor changes in the lamina. The Benettitales, if indeed the constituent families of this group were all closely related, had given rise to the very prominent *Cycadeoidea* group; this appears to have been a most specialised development of the same general flower plan, in that although the pollen-bearing organs were new, the receptacle and ovules differed little from the *Weltrichia* plan. Separate Cretaceous species of both the pollen genera *Classopollis* and *Eucommiidites* occur.

Subsequently in Barremian to Albian time a new polycolpate elongate pollen grain appeared (*Ephedripites*); no plant association is known, and the nearest parallel pollen is of the living *Ephedra* which as a plant could well have been 'fitted in' as a Mesozoic advanced seed-plant. It happens now to flourish as a shrub in semi-arid environments, but its Cretaceous ancestor could well have been a flood-plain or riverine forest plant; there has been enough time for the living genus *Ephedra* to have become a relict in its present specialised environment. The other new advanced seed-plants experimented very tentatively with an expanded leaf lamina not restricted by anastomosis of venation, and could have been named as a group from the character; only because much more elaborate versions of such a leaf have radiated later, these Albian seed-plants have been called angiosperms. The tricolpate-pollen plants have also been called angiosperms, and they may be the same plants as those providing the leaf lamina; as an isolated plant organ, however, *Tricolpites* must have appeared at the time to be simply one more interesting 'experimental type' like *Classopollis*, *Eucommiidites* and *Ephedripites* before it. The newly occurring fruits such as *Onoana* and *Nyssidium* were time-correlated with *Tricolpites* but also with *Ephedripites* and finally with the *Clavatipollenites* group which although monosulcate also possessed a tectate exine; this last was apparently an advanced seed-plant innovation connected with insect pollination and special physiology of presumably a more specific kind.

The plant groups of *Leptostrobus*, *Caytonia*, *Cycadeoidea* (Bennettitales), *Classopollis*, and *Eucommiidites* all became extinct at this time or soon afterwards; this world-wide phenomenon is seldom discussed.

12.6 SUCCESSIVE RADIATIONS

Although by no means proved, and particularly so because palaeolatitudes have not yet been widely used in discussion, there appear to have been successive geographic radiations of certain plant groups followed quickly by their replacement by successors in the low palaeolatitudes.

From mid-Jurassic time Ginkgoales (north and south) and Czekanowskiales (north) became restricted in distribution in this way. From mid-Cretaceous time Pinaceae and Taxodiaceae (north) and Podocarpaceae and Araucariaceae (south) became geographically isolated, perhaps at the same time as they were becoming distinct in some characters from the Mesozoic Brachyphylls and Linearphylls; such necessary constructive evolutionary expressions as 'Protopodocarpaceae' should clearly be defined from fossil information alone, and used both geographically and in a time-restricted sense.

The complementary geographic restriction to low latitudes without appreciable radiation applied first to Nilssoniales and Benettitales and subsequently, to some extent, to Cupressaceae from the Cretaceous onwards.

In Palaeogene time both north and south herbaceous angiosperms became isolated, no doubt with the effect intensified by steady climatic deterioration through the Tertiary and sharpening of latitudinal climatic belt distinctions.

12.7 CRETACEOUS CLIMATIC PATTERN

Although again by no means proved (see chapter 7), the postulate of a time of crustal temperature rise with a maximum (Radmax) in the Campanian–Maestrichtian ages appears to fit all the palaeobotanic and rather diverse palaeozoologic information and to dispose of the Cretaceous–Tertiary boundary faunal 'mysteries'. The scale effect of such change could have been as small as the changes now required for glacial initiation, although a glacial period should perhaps be regarded as only the ultimate manifestation of deviation from average.

The build-up to such a maximum during Cretaceous time could well have provided the great mid-Cretaceous stimulus to seed-plant innovation and sustained it from Cenomanian time onwards. The reversal in earliest Tertiary times seems to have initiated the very large land-plant radiation into temperate latitudes which enabled land vertebrates to dominate once again after a striking hiatus. It does not seem necessary to involve any other non-biologic causes for these Cretaceous biologic events.

12.8 SEED-PLANT ANCESTORS TO ANGIOSPERMS

The general conclusion that the organic evolutionary process continued normally through Cretaceous time does not answer the desire for a direct indication of specific ancestors and for a decision on whether the living angiosperms were monophyletic or not. Unfortunately precise replies to these points are not yet available and it is necessary to accept that the full results will only appear slowly; it is however implicit in the present reasoning that a solution is entirely possible and will follow as the necessary fossils are examined and interpreted.

It is tempting to regard as possible ancestors those groups which became extinct at about the time of the mid-Cretaceous angiosperm beginnings (fig. 12.1). These have all previously been examined

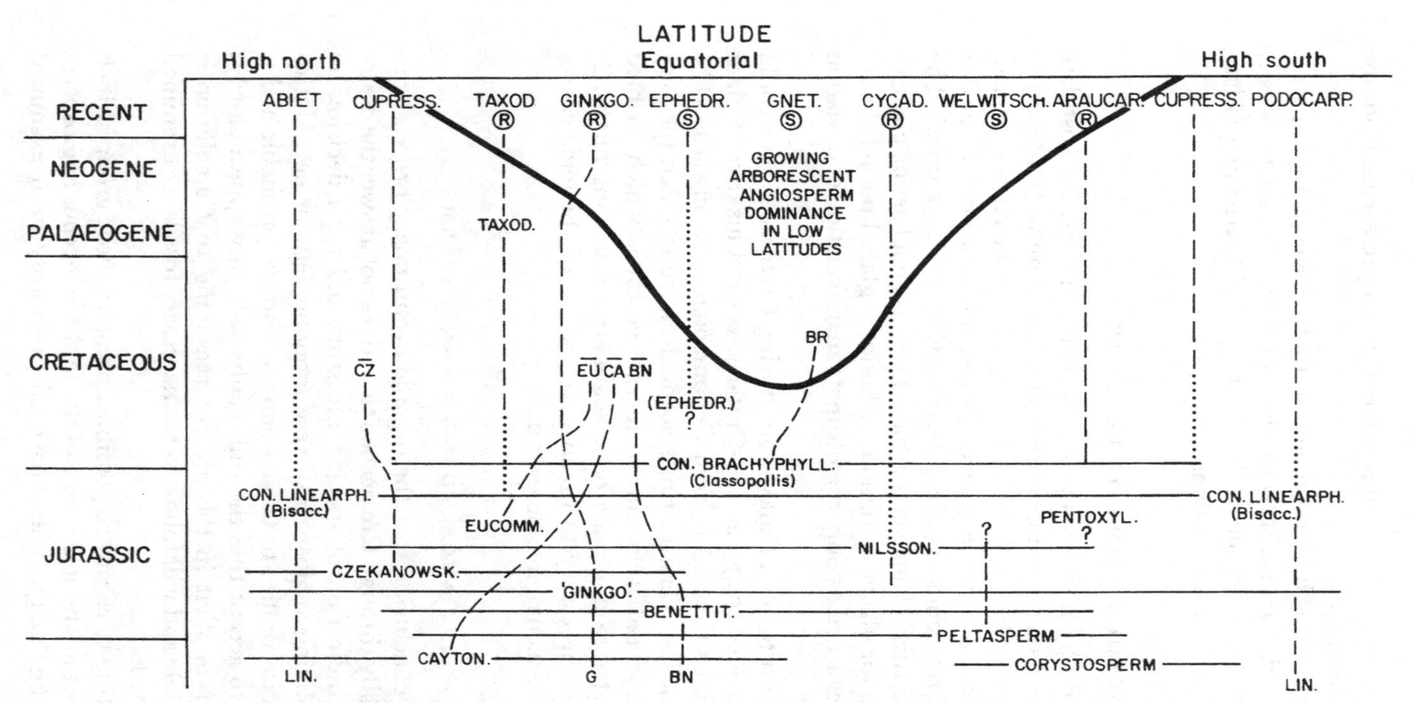

Figure 12.1. Experimental diagram linking the diverse range of Jurassic gymnosperm groups, through Cretaceous and Tertiary time, with the living gymnosperm survivors. Latitude (Recent) and palaeolatitude (Jurassic) are expressed only generally; full line indicates geographical extent only. Broken line indicates probable descent or time-distribution; dotted line indicates less certainty. Supposed early and late Cretaceous presence and distribution of groups is indicated by these lines, and is related to the timing of angiosperm radiation. CON, conifer; G, 'Ginkgo' group; BN, Benettitales; R, relict group; S, small group.

closely with a resulting negative consensus in each of such cases as the Nilssoniales, Benettitales, and Caytoniales. It seems however that this selection is unnecessarily restricted, and that *all* of the Mesozoic seed-plant groups with early Cretaceous representatives contained 'experimental developments' in different organs and functions, and could be eligible; an example of an over-riding difficulty in such matters as the details of the reproductive systems is that although the pollen tubes of a living *Pinus* species may develop unusually slowly, it is difficult to estimate what part, if any, this feature may have played in a Cretaceous pine ancestor. In other words it remains difficult to dissociate from this problem any evidence derived solely from living organisms. At any time, however, a well-petrified fossil may reveal a relevant character and resolve this difficulty.

The generally accepted monophyletic argument is usually based on a feature such as the eight-cell female gametophyte and the associated double fertilisation. There are several points about such an argument: (*a*) no such detail has been seen in pre-Quaternary material; (*b*) it seems quite probable that a female gametophyte reduction of this kind is a logical minimal consequence of full enclosure of the ovule and of the use of a style by pollen-tubes; (*c*) the details of existing female gametophytes are by no means uniform. Such a hypothesis should at least be under suspicion.

A polyphyletic hypothesis has an obvious beginning in the undoubted differences between Liliatae and Magnoliatae, and an immediate consequence that once it is proved the group Angiospermae would have to be finally abandoned in any systematic sense (fig. 12.2). Curiously, one point on which fossils are much more certain than the theorists is that the Liliatae, if they began with the Palmae, appeared much later than Magnoliatae; there also seems to be no definite reason to derive Palmae from any of the Magnoliatae. This point has unfortunately been obscured by the various persistent claims for earlier origin of palms, none of which have been substantiated (see chapter 13). Very recently, Doyle (1973), on the basis of some Aptian monosulcate pollen and some Aptian leaves with apically fusing secondary veins with fine cross-veins, has postulated the presence at that time of early herbaceous monocotyledons (Liliatae); this appears, however, to be a premature slipping back into speculation when many more supporting fossil observations are needed.

Of the many seed-plant groups represented in early Cretaceous

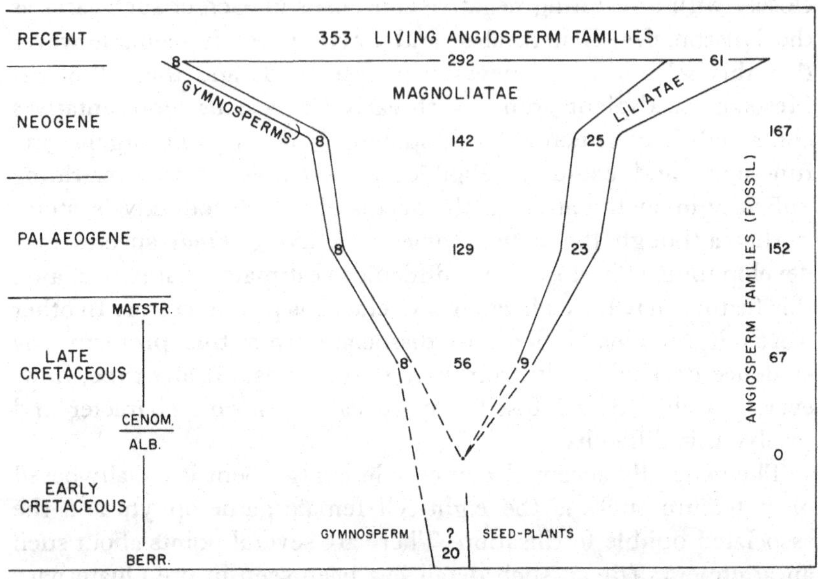

Figure 12.2. Cretaceous and Tertiary angiosperm succession. Fossil family numbers from Chesters *et al.* (1967); living family numbers from Cronquist (1968). Gymnosperm fossil families estimated. The source of Cretaceous angiosperm families is, as noted in the paper concerned, almost certain to be in error in its identifications and its selection; despite this, however, the numbers of fossil families may still be of approximately the correct order.

rocks (chapter 9), several should continue to be considered (fig. 12.3). Krassilov (1973*c*, *d*) favoured consideration of the Czekanowskiales because of supposed stigmatic surfaces in species of *Leptostrobus* and because of the probably-linked very small size (about 250 μm) of their ovules. Thomas (1925) originally described the Jurassic Caytoniales as angiospermous and although this has been shown to be incorrect in the traditional sense, some of the unusual structures of the ovule could have developed further. The reproductive structural detail of both these groups has been taken almost entirely from mid or late Jurassic fossils so that there was time for further development before Barremian time. The Benettitalean flower is well documented into early Cretaceous time in the group *Cycadeoidea*, which, however, has no angiospermid features other than the actual flower; it is not yet clear, because of paucity of fossils, how far the flowers of the Williamsoniaceae had developed in Cretaceous time. The Nilssoniales have been connected so convincingly with the living relict Cycadales that they have not also been

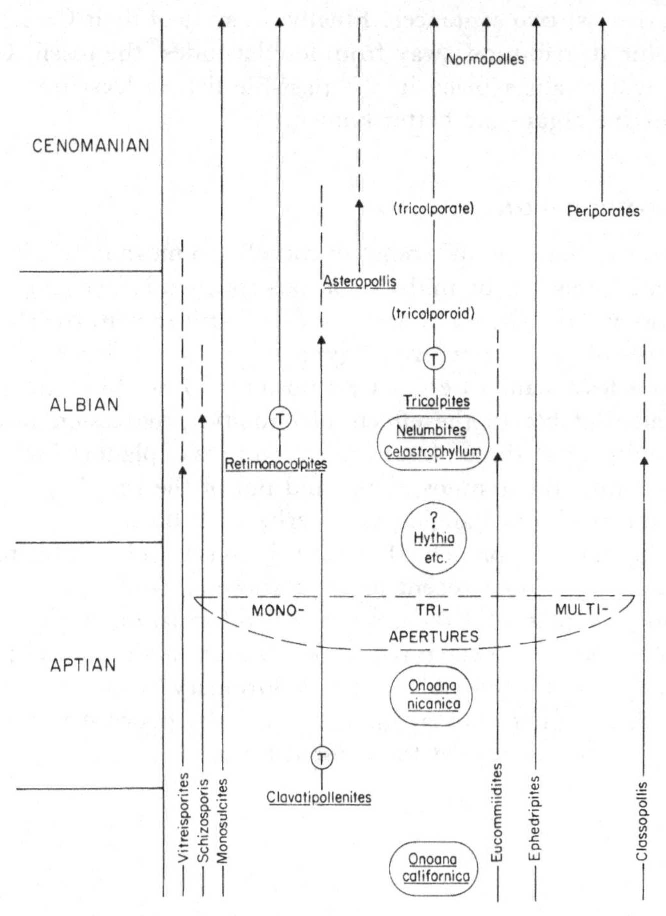

Figure 12.3. Time-distribution of mid-Cretaceous seed-plant pollen fossils, arranged by type of aperture; T indicates tectate pollen. Names of significant megafossil occurrences are ringed.

considered as angiosperm ancestors; there is however some diversity in the fossil members of this small group which is supported by the subsequent differences between the living *Cycas* and *Zamia* groups, although there is a general trend in living species to very large cones in all of them. Such features as bisaccate pollen appear to rule out from consideration the early forerunners of the Pinaceae but the kind of siphonogamy in some living Pinaceae is much more angiospermid than the zooidogamy and haustorial pollen-tubes of Cycadales and Ginkgoales. However it is important to emphasise how easy it is to slip into discussion of Recent rather than Cretaceous plants, as is seen in the last two sentences. Finally, in spite of their Cretaceous geographic distribution away from low latitudes, the fossil 'Ginkgoales' will retain a place in the possible list, at least until their reproductive organs are better known.

12.9 CONCLUSION

If the word 'angiosperm' comes eventually to mean only a level of organisation reached by mid-Cretaceous (or even Palaeogene) seed-plants on several phyletic lines, there will still remain the striking differences of such plants from 'gymnosperms': of thin cuticle, of deciduous leaves and of ephemeral flowers. These differences may all be attributable to the effects of climatic progression towards end-Cretaceous Radmax, but they are features of plants which were woody as were the gymnosperms, and not of the very large subsequent (but irrelevant) angiosperm herbaceous floras.

The suggested approach, therefore, is to set aside all theoretical phylogeny, even such recent ideas as those of Doyle (1973), until a genuine surplus of Cretaceous fossil information has been accumulated and has been stored under entirely neutral descriptors. Then only will it be possible to test evolutionary hypotheses against further data, rather than against other equally poorly-based hypotheses as is the position at the present time.

PART 5

Other theories

13 Pre-Cretaceous angiosperm claims

Since 1900 relatively numerous claims have been made that various fossils of from Carboniferous to Jurassic age were either angiosperms or displayed unequivocal angiospermid characters. Some of these claims have been shown to be simply mistaken, but many concern interesting gymnospermous fossils the more important of which will be illustrated in this chapter. Discussion of many individual cases has already been provided independently by Scott *et al.* (1960) and Hughes (1961*a*); unless new points have been raised subsequently, some of the less important cases of misinterpretation are omitted from this text, although included for completeness in tables 13.1 and 13.2. Some early Cretaceous fossils are included in the tables for continuity but are discussed in the text in chapter 10.

Since 1960 the serious claims have decreased in number and will be taken in some detail. In all cases, those in which the specimens were numerous and well preserved, or in which the observations have been successfully repeated, are accorded the most attention.

All pre-Barremian claims have been carefully considered and for different reasons almost all are rejected; a very few are suspended until the observation is repeated. None is accepted. The general position of argument has therefore changed little in the last decade and a half.

13.1 MEGAFOSSILS

The earliest fossil concerned is *Sanmiguelia* (fig. 13.1) from the Triassic of Colorado, claimed as a palm by Brown (1956); the preservation is a very clear large impression in red siltstone but without organic matter. Becker (1971) mentioned further material, but no cuticle or evidence other than general morphology, in support of this attribution to the palms. *Sanmiguelia* is certainly a hitherto unknown fossil but lacks any exclusively angiosperm character. Two somewhat comparable large leaf fossils are known from

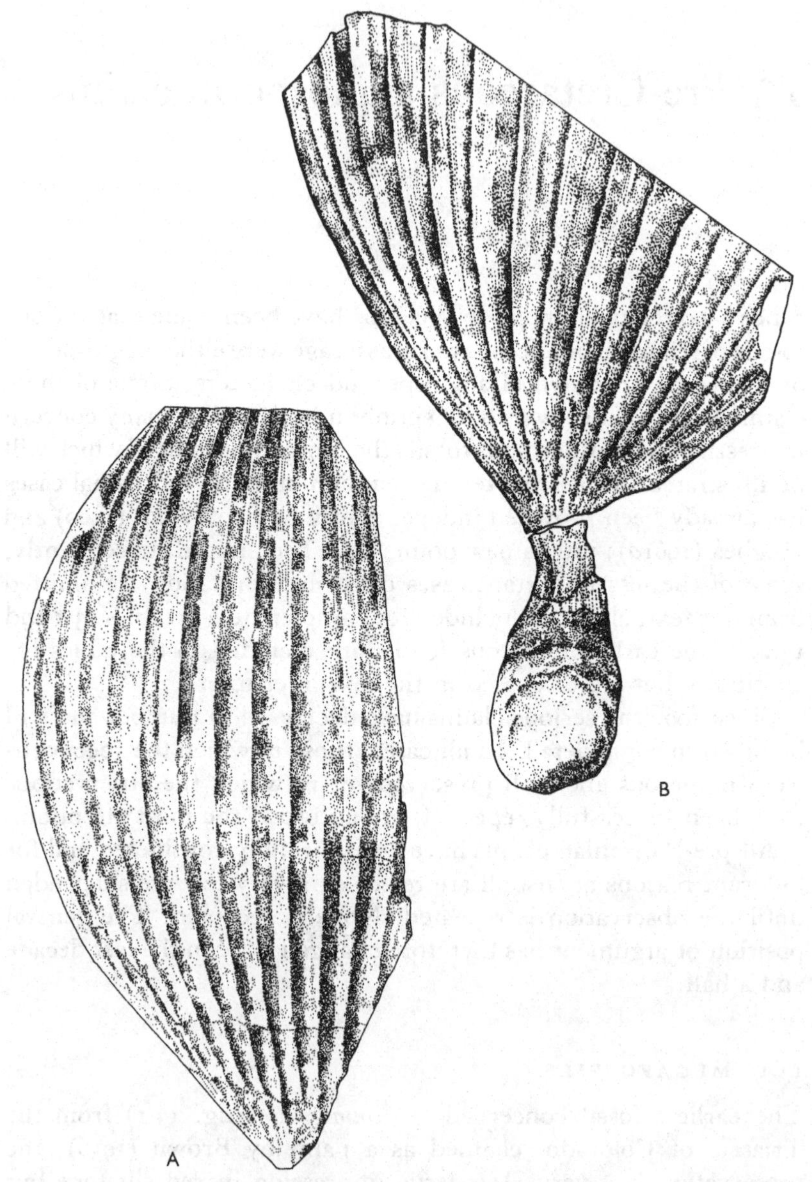

Figure 13.1. *Sanmiguelia lewisi* Brown; leaves in impression preservation; from the late Triassic Dolores Formation, south-western Colorado, USA; ×0.25. A ovate ribbed leaf. B part of leaf attached to stem fragment which is a calcitic 'cast' without visible cell structure. (After Brown 1956.)

Table 13.1. *Pre-Aptian angiosperm megafossil claims*

Age	Taxon	Preservation	Author[a]	Date	Specimen numbers	Remarks
Barremian	*Onoana*	Good	Chandler and Axelrod	1961	1	
Hauterivian						
Valanginian	*Carpolithus*	Fair?	Chandler	1958	1	Specimen lost
Berriasian	Tyrma fruits	Good	Krassilov	1973	Many > 10	
'Tithonian'	*Problematospermum* (+Krassilov 1973)	Good	Turutanova-Ketova	1930	> 1	?Benettite cuticle
Kimmeridgian	*Palmoxylon*	Good	Tidwell *et al.*	1970	Several > 2	Tertiary fossil in mélange
	Ungeria	Poor	Salfeld	1908	1	No detail
Oxfordian	*Sahnioxylon*	Fair	Bose and Sah	1954	1	No vessels
	Montsechia	Poor	Teixeira	1954	2	Indeterminate leaves
Callovian						
Bathonian	*Phyllites*	Poor	Seward	1904	1	No detail of venation
	Sogdiania	Good?	Burakova	1971	1	
Bajocian	*Suevioxylon*	Fair	Kraüsel	1956	1	Withdrawn by author
	Caytonia	Good	Hamshaw-Thomas	1925	Many	Recognised as gymnosperm later
Early Jurassic	*Propalmophyllum*	Poor	Lignier	1895	1	No angiosperm character
	'Graminae'	Poor	Reissinger	1952	Many > 10	Indeterminate
	Sassendorfites	Good?	Kuhn	1955	1	Specimen lost; probably error
	Fraxinopsis	Fair	Wieland	1935	Several	Cycadophyte cone-scales
Late Triassic	*Furcula*	Good	Harris	1932	Many	Probably pteridosperm
	Sanmiguelia	Good	Brown	1956	Several	No angiosperm character

[a] For those papers not quoted in reference list see Hughes 1961*a*.

Table 13.2. *Pre-Aptian angiosperm pollen claims*

Age	Taxon	Author[a]	Date	Specimen numbers	Remarks
Barremian	*Clavatipollenites*	Couper (Kemp)	1958 (1968)	Many	Accepted
Hauterivian					
Valanginian					
Berriasian	*Tricolporopollenites*	Burger	1966	?4	Unsupported observation
Tithonian	*Pterocarya*	Rouse	1959	Many	Interpretation error
	Trifossapollenites	Rouse	1959	Many	= *Eucommiidites*
Kimmeridgian					
	Sporojuglandoidites	Vishnu-Mittre	1955	1	Indeterminate fossil
Oxfordian					
Callovian					
Bathonian	*Magnolia*-type	Simpson	1936	Several	Interpretation error
	Nelumbium-type	Simpson	1936	Several	Interpretation error
Bajocian	'*Clavatipollenites*'	(Tralau)	1968	Several	?Cycadophyte
Early Jurassic	*Eucommiidites*	Erdtman	1948	Many	Gymnospermous
	'*Clavatipollenites*'	(Schulz)	1967	Several	Not this genus
Late Triassic	*Eucommiidites*	(Scheuring)	1970	Several	New species?
Early Triassic					
Permian					
Late Carboniferous					
Early Carboniferous	*Tetraporina* } *Triporina* }	Naumova	1950	Many	Algal aplanospores

[a] For those papers not quoted in reference list see Hughes 1961*a*.

the USSR: *Phylladoderma arberi* Zalessky (presumably a gymnosperm) from the late Permian and *Schizoneura ferghanensis* Kryshtofovich (a pteridophyte) from the Triassic. Although *Sanmiguelia* is distinct, knowledge of these other fossils indicates that the affinity possibilities are wide.

Furcula granulifera (fig. 13.2) was described from the Rhaetian of Greenland by Harris (1932) and was represented by numerous leaf fragments and cuticles with syndetocheile stomata. *F. uralica* Prynada was subsequently described from the USSR and considered to have affinity with pteridosperms such as *Gigantopteris*; in general this is also the view of Scott *et al.* (1960).

On the various pre-Kimmeridgian Jurassic fossils, there are no new observations. *Sogdiania abdita*, however, is a new mid-Jurassic fruit-like body (fig. 13.3 *C*) described from Central Asia by Burakova (1971) and not yet fully interpreted.

The *Palmoxylon* specimens described by Tidwell *et al.* (1970, 1971) from the late Jurassic of Utah have been proved fairly quickly (Scott *et al.* 1972) to be genuine Tertiary palm fossils transferred in a complex local geological situation and not, therefore, of Jurassic age.

Problematospermum ovale was described by Turutanova-Ketova as long ago as 1930 but had been overlooked; ironically this may have been because the author described it as gymnospermous. The specimens (fig. 13.3) came from the famous late Jurassic deposits of Karatau in southern Kazakhstan, and each consisted of a small ovoid-elongate body with a distinct pappus on a tube at one end and an overall measurement of 2.5 cm. Epidermal cell detail has been observed (Krassilov 1973*a*). The obvious outward resemblance is to the achene and pappus of *Taraxacum*, a common living member of the 'advanced' and late-evolving family Compositae not otherwise known before the Tertiary. If the characters of this fossil can be shown to have been concerned with seed distribution, this would certainly be a surprise within the framework of gymnosperms as currently understood; on the other hand, however, small seeds such as those in *Leptostrobus*, were already present in the Jurassic, and the development of hairs for special purposes had progressed considerably by the time of the early Carboniferous ovule *Salpingostoma*. There seems to be good reason to investigate such a fossil further, but no reason to place decision on affinity as a high priority in such investigations.

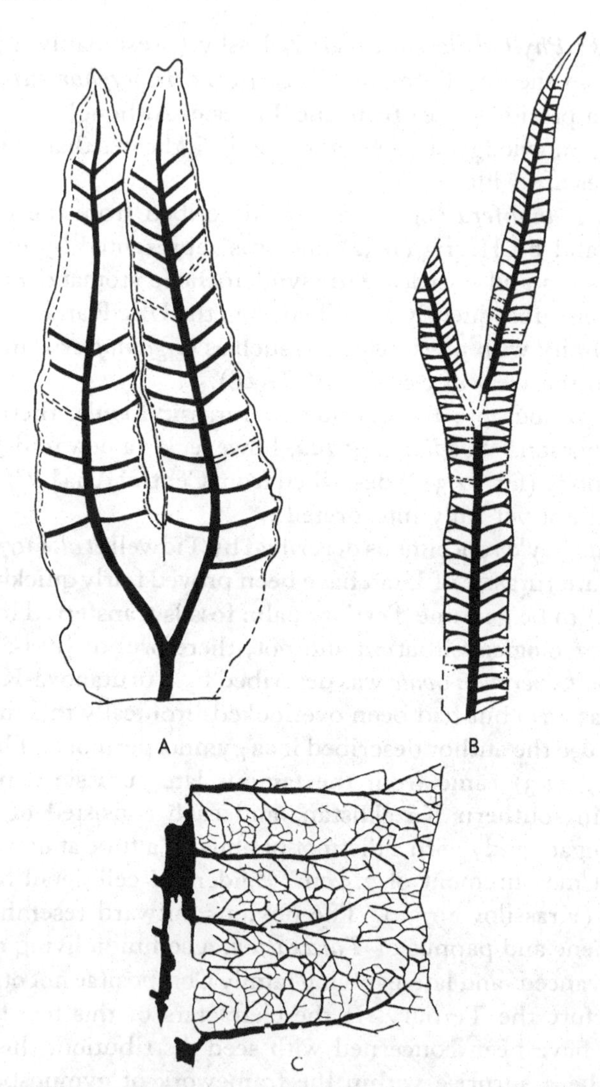

Figure 13.2. Leaves of *Furcula*, late Triassic (Rhaetian) age. A *Furcula uralica* Prynada, ×1. B *F. granulifer* Harris; Rhaetian, East Greenland. C *F. granulifer*; detail of venation, ×5. (A after Orlov 1963; B, C after Harris 1932.)

13.2 DISPERSED POLLEN

The fossils concerned nearly all occur in quantity and as a result have been fully investigated.

Tetraporina Naumova 1950. The miospores attributed to this genus from the early Carboniferous of the Russian platform were

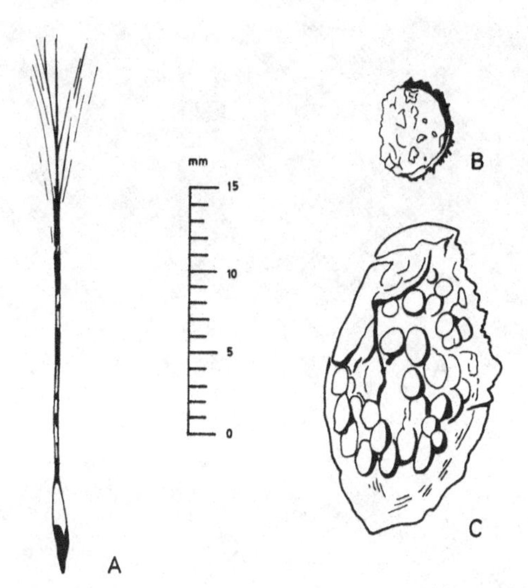

Figure 13.3. Jurassic 'fruit' and seed fossils; ×4. A *Problematospermum ovale* Turutanova-Ketova 1930; from late Jurassic of Karatau, south Kazakhstan, USSR; after Krassilov (1973*a*). B fruit-like capsules from Tyrma River, Far East USSR; age Tithonian–Berriasian; after Krassilov (1973*c*). C *Sogdiania abdita* Burakova 1971; from middle Jurassic of Central Asia; after Burakova.

still described as angiosperm pollen in 1958 by Teterjuk; they have subsequently been shown to be algal aplanospores as suggested by Scott *et al.* (1960). They have also been seen outside the USSR and illustrated by Playford (1963) from the early Carboniferous of Spitsbergen.

Eucommiidites Erdtman 1948. The original species *E. troedssonii* described from the early Jurassic of southern Sweden was of medium size (30–40 μm) and oval with one prominent deep furrow or colpus and two less distinct ones (figs. 13.4, 13.5). In the rarely seen end views the appearance is of a tricolpate pollen grain. Several species have been described and the range of the genus is from late Triassic (Switzerland: Scheuring 1970) to mid-Cretaceous. Erdtman originally suggested affinity with the pollen of the dicotyledonous tree *Eucommia* which has unequal colpi; Couper (1956, 1958) showed from its symmetry and orientation behaviour that it was more probably gymnospermous. Hughes (1961*b*) described the Jurassic Bathonian pollen as distally monocolpate and proximally zonosulcate; he also demonstrated the gymnospermous origin by describing specimens of another (Cretaceous) species lying in the pollen chamber of an ovule with a very long micropyle. This last observa-

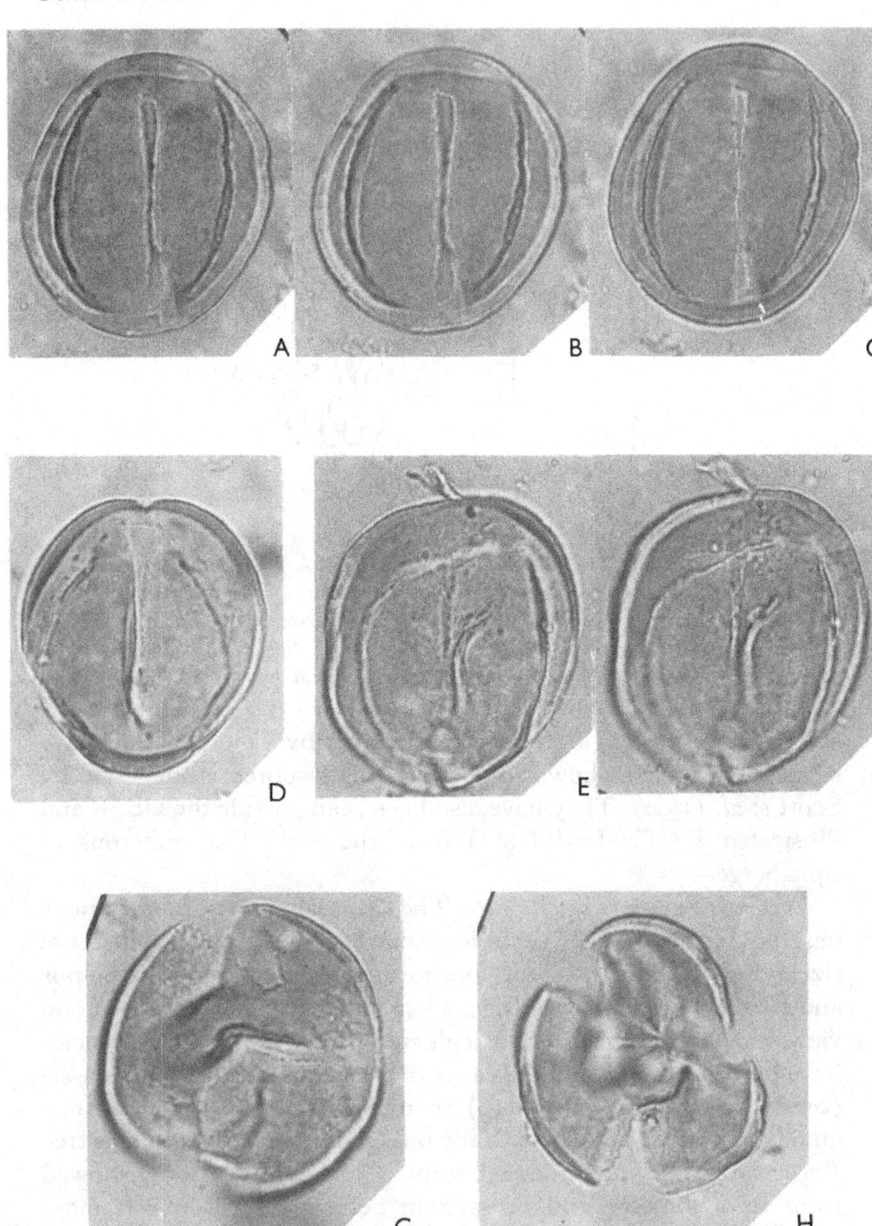

Figure 13.4. *Eucommiidites troedssonii* Erdtman; preparation C104/1 of sample B13, Brora Coal, Sutherland; mid Jurassic; ×1000. A–C back view: A high focus; B mid focus; C low focus. D slightly oblique front view. E, F slightly oblique back view: E low focus; F high focus. G end view, inflated specimen; main sulcus right. H end view, partly inflated specimen; main sulcus right.

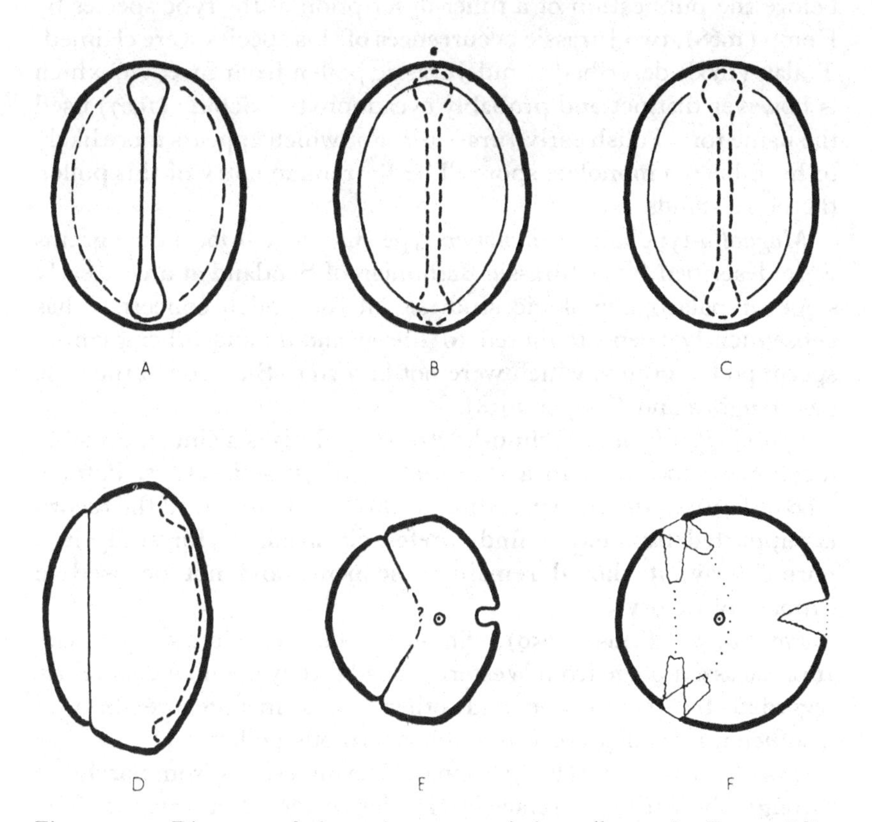

Figure 13.5. Diagrams of the main aspects of the pollen grain *Eucommiidites troedssonii* Erdtman; for comparison with fig. 13.4, approximately ×1000. A front view showing main sulcus, with ring sulcus (broken line) behind. B back view with complete ring sulcus. C back view with incomplete ring sulcus. D lateral view showing different convexity of front and back faces. E end view (hypothetical); unexpanded furrows. F end view, expanded grain; main furrow right.

tion has been subsequently repeated by Brenner (1967*b*) on material from the early Cretaceous of eastern North America, and by Reymanovna (1968) from the mid-Jurassic of Poland. Although this pollen was clearly gymnospermous, it is unlike all other gymnosperm pollen except perhaps *Classopollis* in having a symmetry which is unaccounted for in purpose outside the angiosperms. It suggests indeed that the very distinction of the plant groups under scrutiny is for the Cretaceous period perhaps wrongly stated.

Clavatipollenites Couper 1958. This Cretaceous Barremian pollen appears from morphology to be angiospermous. Unfortunately just before the publication of a fuller description of the type species by Kemp (1968), two Jurassic occurrences of this species were claimed. Tralau (1968) described a mid-Jurassic pollen from Sweden, which is however distinct and probably cycadophytic. Schulz (1967) used the name for a Polish early Jurassic grain, which appears more likely to have been a monolete spore. The Barremian entry of this pollen therefore stands.

Magnolia-type and *Nelumbium*-type Simpson 1936. These grains were described from Jurassic Bathonian of Scotland at a very early stage of palaeopalynologic studies. All the pollen concerned has subsequently been attributed to *Eucommiidites* and other gymnosperm pollen groups which were not known to Simpson at the time (see Hughes and Couper 1958).

Sporojuglandoidites Vishnu-Mittre 1955. This is a single, possibly incomplete specimen in a thin section of Jurassic chert. Potonié (1960) doubted the interpretation of this fossil, and until the record is supported by another find, preferably available for study in a normal way, it should remain in suspense and not be used in support of theory.

Pterocarya (Rouse 1959). These grains were found in a late Jurassic assemblage from Western Canada; they are now apparently regarded by the author and others as a misinterpretation of insufficiently well-preserved gymnospermous pollen.

Tricolporopollenites (Burger 1966). The material is from boreholes through the earliest Cretaceous in the eastern Netherlands. The number of specimens is not stated in the text, but appears from the tables in the paper to be four, from three samples in two boreholes. Spores of *Cicatricosisporites* described from the same material indicate on interpretation a Berriasian age. The single figured pollen grain appears to be what is implied by the name *T. distinctus* Groot and Penny 1960, of which the type material is of late Cretaceous age and is by no means primitive when compared with the main Albian and Cenomanian records. It so happens that extensive studies of rocks of this Berriasian age in East Germany (Döring 1965) and in southern England (Hughes and Moody-Stuart 1969, Batten 1973, Hughes and Croxton 1973) have not revealed any comparable pollen. Although it may appear to be slightly unappreciative of Burger's record, a repeated observation is considered necessary before altering theory on account of this one record. For comparison, the original description of *Clavatipollenites* by Couper (1958) was based on only five specimens, but the description was fully secured by Kemp (1968) using more than a hundred specimens all obtained by painstaking study of the sample from which the holotype had been taken by Couper.

13.3 CONCLUSIONS

It inevitably appears to suggest lack of appreciation of other work if one dismisses so many records made over a long period of time. The dismissals are, however, made individually and are in almost all cases of the interpretation rather than of the observation.

It could be argued that in each one of these claims the basis of interpretation is an observed character which is judged to parallel a character of a selected living angiosperm plant. In addition, the majority of the claims have been made since 1945 in a period when general evolution theory has highlighted the supposed inadequacy of the fossil record, and therefore the possibility of an unseen upland flora; this has made common the rapid acceptance of a single character as a guide into the unknown. As a result all the cases have tended to fail to similar investigation and in the same way; this was not anticipated at the start of this investigation.

In all cases, however, aspects of sedimentation, of diagenesis and preservation, and of statistical significance of the observations, have been considered quite separately.

The conclusion is that none of these pre-Cretaceous fossils represents an angiosperm. In pre-Barremian Cretaceous time the position is, as expected, less clear-cut, and such fossils as the Tyrma fruits (Krassilov 1973*c*) may possibly represent the beginning of the introduction of new significant and angiospermid characters into the early Cretaceous gymnosperm flora.

14 Contribution of studies of comparative morphology

Studies of comparative morphology of living plants, although obviously a valuable discipline in themselves, have also been applied to the elucidation of two major problems: first they have been directed to the classification of living angiosperms and other plants, which is a legitimate use; secondly they have frequently been quoted as a basis for phylogeny in evolution, which is a false use.

This second use, as a substitute for genuine historical information derived from fossil evidence, is false in the sense that it may offer a guide to phylogeny in a very crude and general way but it cannot alone lead to any kind of detailed solution. As an assault on the problem it is comparable with the well-known 'armchair' (or in the field, 'binocular') geology; it is always possible to make some progress in this way but nothing further will result without hammering and measuring the rocks in question, which activity seldom fails to produce a profound effect on any theory.

14.1 PRESENT APPLICATION TO PHYLOGENY

For over a century but particularly in the last fifty years many competent and distinguished botanists have used comparative morphology of extant plants in attempting to reconstruct both a primitive flower and more recently other characters of an ancestral plant. This has resulted in many ingenious theories which cannot be tested or which lack any kind of response in the fossil record. Recently, however, a series of three brief and elegant papers by Sporne (1969, 1970, 1972) has provided a statistically-based cap for this kind of work in the form of his 'advancement index' for families of dicotyledons; these useful papers in effect summarise a large amount of work conveniently and are thus central. A fourth paper on 'The survival of archaic dicotyledons in tropical rain-forests' (Sporne 1973) is apparently different in that some fossil evidence is drawn from Muller (1970) and in a more qualified way from Chesters *et al.* (1967);

the conclusions of that particular paper read as follows: '(*a*) Twelve characters are more abundant among rain-forest families of dicotyledons than they are in the flora of the world. (*b*) Nine of these were more abundant in pre-Tertiary or pre-Oligocene times than they are at the present day. (*c*) It is argued therefore that these characters are archaic survivals in present-day rain-forests.'

Such conclusions are bland enough to make any lack of acceptance sound perverse, particularly as (*c*) must have at least an element of truth. The difficulties are those items that are only implicit in the reasoning, the logical consequences, and the choices open for further study.

One or two of the steps in this kind of reasoning derive from statements in Sporne's paper but most of them have been buried in the widely scattered work on which the paper is built; identification and discussion for each of the selected steps are as follows:

1. That an extant family can be described as possessing certain morphocharacters, and that any taxa within the family which do not bear these characters may be regarded as less significant than the majority.
Perhaps the selection of the family to use in this context is neither free nor entirely appropriate; the strength of the data base is as much in doubt as the ecologic and geographic homogeneity of the family.
2. That a few characters such as 'woody habit', 'scalariform features in vessels', and 'pauci-aperturate pollen' (Sporne 1974) are key primitive characters and may positively be expected to have occurred in the earliest angiosperms.
Living gymnosperms certainly bear two of these characters but perhaps as so few restorations from fossils have been made some doubt may remain about most of the characters of both Cretaceous gymnosperms and angiosperms.
3. That the other detailed characters for which a strong positive correlation with these key characters has been shown are also to be expected in the earliest angiosperms.
This is not independent evidence.
4. That certain extant families each bearing a number of these 'primitive' characters are themselves primitive families and that these may be confidently expected to have had a Cretaceous record.
Only one-third of these families with an 'advancement index' below 45 do at present have a pre-Miocene record even claimed, but it

is fair to point out that the other two-thirds are all very small families.

5. That a very small number of specimens of a single fossil plant organ of Cretaceous age may be placed in an extant family without necessarily finding any other organs (of the plant) and without establishing any proof of the continuity of existence of the family thereafter.

The value of such attributions based only on a few leaves or on a few pollen grains is most uncertain.

6. That the Cretaceous existence of a family having been established on such a base, other characters known in the extant members of the family are also presumed to have been present in the Cretaceous.

This is perhaps the most dangerous of all these steps but very difficult to eradicate when the family name has been used for the fossil material.

7. That certain such families persisted from Cretaceous to Recent times without appreciable evolutionary change of characters.

When associated with the last step, this is without basis and could in any case be regarded as leading to circular argument; in almost any other sector of the fossil record such a suggestion would be considered preposterous.

8. That certain lowland tropical ecologic distribution centres acted as 'museums' (in the sense of Stebbins 1972) in making possible the conservation of such families.

Such explanations only add an apparently sophisticated gloss to an extremely tenuous argument.

This selection of points is intended only to show how little attention has been paid to the background purpose of such work, particularly in the lack of allowance for the time-sequence element. Clearly 'classification' was the first object of these studies, and as applied to living plants only and in the narrowest meaning of classification on one time-plane there can be no objection or criticism. The extension into the historical dimension of phylogeny, and the widespread view that classification without phylogenetic control is empty and perhaps almost valueless, have probably led to the present situation. What was until recently almost a palaeontological vacuum, in which could persist the 'mystery' cult concerning angiosperm origin and history, probably accounts for the phenomenon of continued widespread acceptance of these beliefs. They have been widely tolerated by editors and by the scientific public, as is illustrated by the 'phylogenetic' title and the text of a recent review by Sastri (1969), indicating clearly the general attitude in this misdirected literature.

14.2 ACTUOPALYNOLOGY

With the advent of relatively easy application of SEM and TEM (transmission electron microscope) studies to supplement the existing optical microscopic work, there has recently been a widespread extension, in the attractive field of actuopalynology, of the habit of deriving 'evolutionary' conclusions from comparative morphology. Rather curiously there had been a delay in this field caused by an almost universal reluctance to accept 'adaptive significance' of detailed characters of surface sculpture, of infra-tectal sexinous structure, and of wall-layer variety around apertures. Now that an ontogenetic sequence of wall-layer development is known in some cases, and a physiological (or biochemical) explanation of inter-bacula space as storage area for recognition or compatibility substances has been offered (Heslop-Harrison 1975, in press), perhaps all features of pollen walls can be accepted as significant in natural selection despite their nanno-dimensions.

The studies made recently in the pollen of numerous families of living plants show that once again as in the case of earlier experience with gross morphology, a great variety of pollen exine complexity and of aperture arrangement is a normal expectation in the average Recent plant family. Although this has made it very difficult to provide the desired sequences of families from 'primitive' to 'advanced', it has led to extensive evolutionary speculations within most of the families; some of these explanations even involve 'reduction' to account for an apparently primitive pollen in an otherwise advanced taxon. There is, of course, no evolutionary value to this work as it stands.

The plain description of the material taken from the Recent time-plane is valuable and even essential for morphologic interpretation, but evolutionary studies are only directly advanced when the fossil pollen is studied at the same time (Muller 1970, 1974); even then, however, the use of genera and superior taxa should be modified as suggested in chapter 16, to avoid the circular argument.

14.3 CHEMOSYSTEMATICS

Unfortunately, recent work on chemotaxonomy (e.g. Bate-Smith 1968, Kubitzki 1969) and other phytochemical topics (see Harborne 1969) seems bound to follow in the track of the morphologic work and thus be open to similar criticism concerning method. The same applies to the use of chromosome numbers (Ehrendorfer *et al.* 1968). Neither can truly be said to lead to phylogeny, although they certainly provide refinement of classification possibilities when confined to the one time-plane. It can be argued that chromosome information and phytochemical data are primary and more powerful than the traditional morphologic characters and that therefore phylogenetic proposals based on them are of much greater and perhaps even adequate value. It is tempting to accept this, but it is almost certainly illogical because these new characters are not independent of the old and are not fundamentally different. The interpretations of such new facts are still in their early stages; it seems possible that the current chromosome counting may prove to be too crude; on the chemical side, the development of protective alkaloids is probably an advanced feature of plant–animal integration and the loss by natural selection of certain chemical characters seems equally probable. The true time-dimension is lacking and it is not clear yet how or if such chemical and genetic facts may eventually be extracted from fossils.

14.4 FUTURE REQUIREMENTS

Even if the conclusion for the present is that not enough information, morphologic, chemical or genetic, can be extracted from fossils, it is still essential for progress to include the conceptual possibility of such information if any time-phylogeny is the desired aim. In view of recent achievements in these fields it does not appear too unrealistic to plan in this way. In comparison with the parallel but much more intractable (because of less potential characters) field of algal phylogeny (see Frederick 1970) this may appear reasonable.

Until characters of all kinds are extracted from fossils of single specified geological time-divisions, and are handled quite separately from those of any other time-divisions including the present day as one such division, no genuine progress towards time-phylogeny will be made. Work on the present lines that is referred to in section 14.1 above will obviously continue because there are currently so many practitioners but its effect will be comparable to that of studying the moon or Mars before any instrument-landing had been achieved. The 'landing' in the case of angiosperms phylogeny is a matter for palaeontologic skill and persistence as much as for financial investment; it could be speeded if all workers on Recent pollen would explore the time-range of their taxa, however short these may prove to be and however apparently unsatisfactory study of dispersed fossil pollen may be.

15 Other current theories

The advocacy of a new approach obviously must imply the rejection of numerous published previous theories as inadequate, although in many cases ideas which are not necessarily central to the theory concerned are valuable and well worth re-incorporating. A brief account is therefore presented of a selection of earlier or current theories with this retrieval aspect in mind.

Botanical opinion on the unique and homogeneous nature of the qualities of the angiosperms has been so strong that with one exception all theories have been based on a monophyletic origin for the group or at most an extremely restrained modification of this.

15.1 CATASTROPHIC THEORIES

Relatively simple explanations of a supposedly sudden angiosperm origin have attracted geologists. Earliest was perhaps the invocation of a universal Albian–Cenomanian marine transgression as a stimulant of some sort, rather reminiscent of biblical inundation. The 'transgression' concerned was almost certainly an epeirogenic immersion and happened to be most marked where is was observed in France and in North Africa on either side of Tethys; in many other parts of the world there was no sign of it. At most, therefore, it was a regional phenomenon and not relevant to a world-wide palaeobiologic problem.

Teslenko (1967) suggested a decrease in atmospheric carbon dioxide but there is no other evidence of this, either physical or biological. Axelrod (1967, 1972) suggested drought as a stimulus although it could not have been universal; on the whole the atmosphere–hydrosphere system appears to be large enough for rapid and reversible simple changes of this kind to be very unlikely.

Admittedly the temperature maximum (Radmax) suggested above in chapter 13 might appear to belong to this category, but in that case the rise and fall of temperature were spread over at least two geologic periods and many other phenomena appear to fit in.

Unfortunately it is also necessary to disagree with various Russian friends (e.g. Vakhrameev 1972) and with Frederiksen (1972) who refer to a 'Mesophytic' era from early Permian to mid-Cretaceous as being marked at beginning and end by unspecified natural events which changed the land-flora. Although this disagreement appears also to challenge the use of the word 'Mesozoic', both are unnecessary terminology and merely record a rather local view of events and the plant term adds nothing to understanding at the universal level.

15.2 FOSSIL RECORD FAILURES

The most persistent negative contribution in this category is to the effect that all complex organisms took a long time to evolve, that early stages might have been rare, and that they were therefore statistically unlikely to occur as fossils. While there is much evidence that radiations of organisms took place very rapidly in geological time, for example in ammonoids on several occasions, and in some mammal groups, it is unhelpful to burden investigation with such negative theory except as a last resort or admission of failure.

The 'upland' theory of angiosperm origin originates partly from 'length of time necessary to evolve' and partly from a kind of botanical 'uniformitarianism' which interprets the mode of life of past organisms directly from that of their present descendants. Geologically the probability of preservation of fossils of different kinds from upland or montane organisms in the different successive geological periods can be assessed; complete upland lake deposits or even parts of them are unlikely to survive erosional cycles and the Tertiary examples of such deposits in western North America may well be the oldest to have survived; palynomorphs and other small fossils of montane origin obviously can be preserved by transport into lowland deposits but what are needed are statistical analyses of distribution patterns to demonstrate that the fossils concerned originated outside the deposition area and may be presumed to have come from upland taxa. This is one of the rare instances in which palaeopalynology can be provided with a technique from Quaternary studies, but it has not yet been seriously attempted; hitherto the montane origin of dispersed microfossils has been based on botanical assertions of the 'uniformitarian' kind which should not be accepted without other support.

Other theories

On the general botanical side there appears to be a tendency to regard all land habitats as of equal status, with interfaces through which there is partial flow of taxa, or at least of organisms, in both directions. Historically it appears much more likely that the true position is one of a crowded lowland tropical pool of organisms in which the chief control of natural selection is biological, and radiations into high altitudes and high latitudes where the principal control is physical. The possibility of migration back from upland to lowland is apparently assumed in 'upland origin' theory and is necessary to the maintenance of such theory; no such flow has been demonstrated nor any detailed mechanism suggested, and this lack appears to be a very serious flaw. With marine life a similar situation exists in the relationship between equatorial shelf seas and the rest of the oceans; the latitudinal gradient of decreasing numbers of taxa (but not of organisms) towards the poles has been used with some confidence in prediction by marine palaeontologists. The marine situation is complex in different ways with respect to nekton, which parallels birds; however migration in time of benthonic organisms back towards the equator has not been contemplated or postulated.

15.3 THEORIES OF PURELY BIOLOGIC ORIGIN

Various authors have focussed attention on unusual and isolated fossil gymnosperm groups as possible angiosperm ancestors on the basis of comparison of selected characters, but usually without taking into account their many other characters or their time-distribution. For example Melville (1962) announced a new theory of the angiosperm flower based on a new morphological unit, the gonophyll; almost immediately he found a resemblance to this unit in descriptions of *Glossopteris* fructifications from South Africa, although the fossil material was of impressions only and Permian in age. There were too many uncertainties of interpretation of the fossils for this theory to be accepted widely but in 1969 Melville went further in comparing well-known *Glossopteris* leaf venation details with these observed in a few of the more 'primitive' living ranalean angiosperms. Alvin and Chaloner (1970) in reply pointed out that such dichotomising venation patterns were known from several other groups of fossil gymnosperms in addition to the Glossopteridae; they supported Melville, as apparently do most other botanists, in postulating a single evolutionary origin for the angiosperms and they suggest from the evidence that net venation will have developed within the group during Mesozoic time rather than in the Permian as had been implied. The situation therefore remains very open especially if a polyphyletic origin is to be considered. Another less developed example is the suggestion (Maekawa 1962) that the Russian Permian genus *Vojnovskaya* (see Zimina 1967), which has a bisexual strobilus, may be an angiosperm ancestor; this appears to be a weak case but it is interesting from the point of view of plant–insect integration that this group preceded the Benettitales with the bisexual strobilus.

Meeuse (1970) considered that his previously stated preference for a multiple origin of the angiosperms was greatly strengthened by recently published phytochemical data; his discussion does not attempt to deal with fossils in any detail but shows that even when discussing only living taxa such polyphyletic conclusions may be reached. Meeuse (1961) did earlier suggest the Pentoxylales as a possible origin for monocotyledonous angiosperms and it is understandable that all bizarre fossil groups not apparently connected with the main stream should be tried out; to proceed with such a theory, however, the next need is extension in time and space of the present very limited knowledge of the fossil group.

15.4 COMPOSITE THEORIES

One of the most thorough general accounts of this kind was presented by Nemejc (1956) and although his plan may no longer be acceptable, his data compilation is very valuable. The most comprehensive recent theory is presented by Taktadjhan (1969) in the very useful English edition of his book; he discusses the comparative morphology evidence more than the actual fossils and supports an upland origin in early Jurassic time. There is, however, very little discussion of any critical early Cretaceous fossils.

Krassilov (1973*d*) considered the late Jurassic and early Cretaceous flora of East Asia and interpreted changes of the conifer and Ginkgoales proportions in the floras as climatic deteriorations in mid-Jurassic and Barremian–Aptian times; in reproductive structures he favoured *Leptostrobus* of the Czekanowskiales as being close to angiosperms; in leaves he compared *Proteophyllum* of the Cenomanian of Bohemia with *Scoresbya* of the early Jurassic flora of Greenland. His analysis aims to narrow the gap and wisely discusses only Mesozoic fossils, although the leaf comparison appears to be unpromising with *Scoresbya* too far separated in time to be relevant as an evolutionary succession.

Ever since continental drift was first suggested by Wegener, biologists have been eager to support or disprove the idea by comparing distributions of plants and animals particularly on sections of the southern continents. Although the situation is now very different as a result of knowledge of Mesozoic and Tertiary plate movements, many biologists appear to continue as if there was still the same debate. Schlinger (1974) referenced many such papers in dealing with *Nothofagus* and its associated phytophagous aphids and parasitoids in South America and Australasia; although he dealt with the latest publications on relevant plate movements, he still suggested (1974, p. 338) that entomologists should from this kind of evidence indicate alternative sites for plates. As the ocean floor magnetic anomaly pattern becomes clearer there is less and less scope for this kind of discussion about plate positions in post-Jurassic time, and the biologists concerned would be better employed settling down with the help of the stratigraphic record to plot possible extension and migration routes without disputing the palaeolatitudes further. As is so often done with similar studies, Schlinger postulated an Upper Jurassic to Lower Cretaceous origin and distribution for *Nothofagus* based solely on supposed continental connections and barriers; there is no fossil evidence whatever for such an unlikely postulate. His first evidence was end-Cretaceous dispersed pollen, followed by a more certainly known Oligocene pollen distribution; there are as yet no relevant insect fossils in the area. Such studies of which this one is only taken as a current example can only profitably be pursued step by step back from Recent through late Tertiary time.

15.5 CENTRE OF ORIGIN

Throughout the history of discussion of this problem there appears always to have been a desire to identify a centre of origin and dispersal for the angiosperm group. The Arctic was chosen by Heer (1868) and was favoured by Seward (1933) but this support was based on the wrong dating of some Greenland rocks as early Cretaceous; they have subsequently been shown to be late Cretaceous. Much more persistent has been the assumption that South-east Asia was the cradle of the group (Reid and Chandler 1933, Takhtadjan 1969, Muller 1974). This is mainly because the area now is uniquely rich in 'primitive' angiosperm groups and has relicts in several other plant groups; it also appears (Muller 1970) that the local plant evolution has been continuous since late Cretaceous time. Takhtadjan (1969, p. 156) considered carefully whether the area was simply a 'refugium' rather than a centre of origin; he rejected this idea because tropical Africa and central America were so devoid of the 'primitive' angiosperm groups involved. However, a glance at the maps of crustal plate movements since mid-Cretaceous time explains that these other areas were 'misplaced' at the critical times.

Although Axelrod (1959) has not proved his low to high latitude migration, such an explanation seems much more probable than any centre of origin. South-east Asia with its unique extant flora is probably a 'refugium' for the survivors of what may well have been a general North Tethys early Tertiary (and perhaps latest Cretaceous) angiosperm flora; this flora would have been pressed southwards in Eurasia by the steady climatic deterioration through Tertiary time, being unable in most palaeolongitudes to inhabit suitable ground for survival.

15.6 CONCLUSION

Quite independently of any arguments about the validity of using non-evolutionary comparisons with extant taxa, probably all of these theories fail because they do not (and in most cases could not have done when written) allow for land-area movements now more or less established in plate tectonic theory. They also fail in many cases because so little fossil information has been used.

PART 6

Consequences

16 Classification of angiosperms

Approximately 300000 species of extant angiosperms have been classified into anything from 350 families (Cronquist 1968) to 440 families (Takhtadjan 1969) of which about one-fifth are known as monocotyledons or Liliopsida, and the rest as Magnoliopsida. Classifications of the last century, such as that of Bentham and Hooker which mentioned only 200 families, could not include the many taxa which had not then been identified or even discovered. There is a surprisingly large number of cases in *all* these classifications in which the content of a taxon or its hierarchical position is in doubt, and it is probably still true to say that knowledge of fossils has so far contributed nothing whatever to these classifications. Authors discussing this phenomenon (e.g. Davis and Heywood 1966) point out that the exclusion of fossils has been deliberate because their evidence was too difficult to interpret, and that as the object of classification was purely practical it was better not even to attempt to use a phylogenetic basis.

Judging from the extent of the relevant publications, improvement of the classification has been slow and of uncertain value. It appears to be a reasonable conclusion that until phylogenetic classification is ultimately achieved, the matter cannot be satisfactorily settled or even be made strikingly more efficient by any other means.

16.1 PLANNING OF A PHYLOGENETIC CLASSIFICATION

Because only characters which can be observed in both living and fossil angiosperms will be usable, all characters employed will have to come from those parts of the leaf, wood, pollen and fruit that can be preserved in fossils. Progressively through Neogene time an adequate number of other organs including parts of flowers may be available for supplementation, but in earlier periods the restriction in the last sentence will be absolute and its implications are that no characters of cotyledons, roots, stem apices, nodal anatomy, inflorescences and flowers, will be applicable at any level before Miocene time.

This concept is so different from any basis of classification in current use that its acceptance could only be expected to be gradual. Wolfe (1973, p. 336) severely criticised megafossil palaeobotanists for the lack of progress but was inclined to suggest that palynologists had the key; in this he did not go nearly far enough, because the latter on their own are as unlikely to succeed as were the leaf-palaeobotanists, without a breakthrough of method. Such a method change would affect the whole of angiosperm palaeobotany and would certainly not be understood or tolerated if it appeared to conflict with the *practical* uses of the present classification.

It would be most advantageous therefore if the new scheme were to be introduced first in the Cretaceous only, and were made as rigorous and far-sighted as possible before any contact was made with the 'extant' classification which it would eventually replace because of its phylogenetic origin.

16.2 CHOICE OF FUNDAMENTAL TAXA

In an open situation with only fossil material concerned it is still debatable whether it would be better to erect and use taxa of 'species' rank or to work entirely with characters (see chapter 3). However, as the ultimate object here is to lead to a successful classification of living organisms which are likely to remain arranged in definable 'species', the difficulties with the fossils will be better resolved through the medium of purely artificial taxa erected for the separate leaves, pollen, or other organs. Such organ-taxa will probably be best developed on the lines of the almost neutral palynologic biorecords and comparison records (chapter 4), although an extra mode of expression of a degree of association of the organs will be necessary. This 'degree of association' of fossil organs has not yet been much explored, for whereas occasionally it is possible to relate leaves and fruits by cuticle characters, it is not normally possible to relate leaves to wood directly, or pollen to fruits. Ultimately chemical relation of such data may be available but, for the present, patterns of occurrence in rocks will probably have to suffice; the placing of some numerical or other value on such associations will be required before these fundamental taxa can be meaningfully grouped in time or space.

16.3 HIGHER TAXA

In the handling of extant organisms, higher taxa are simply a filing and retrieval device, or a language of communication, for use on the single time-plane. They have acquired in many animal groups a supposed phylogenetic significance, but in most plant groups and particularly in the angiosperms this is clearly of no such value as no fossils at all have been used.

The classificatory significance of a genus or higher taxon lies really in its *delimitation from others of the same rank* and hence its position in the whole field of such taxa; this statement is believed to be true despite the current nomenclature rules which insist on typification by selection of a cluster 'centre' (type species, type genus) rather than on definition of limits (see chapter 2). Such discussion very clearly relates to the one (present) time-plane.

If, however, a genus or higher taxon-concept is used to include 'species' from a range of geologic time-planes, its significance in terms of limits against other taxa (or even in terms of cluster centres) will be different on each time-plane studied depending on the general state of biological evolution at each of these times. This is not a measure of the number of fundamental taxa in each higher taxon-concept which may either increase or decrease with time, but a statement of the position of the family (or order) in the context of all those other families (or orders) existing at the time selected. This position is seldom expressed, but can be taken from such compilations as Harland *et al.* (1967).

From this it follows, as indicated in fig. 16.1, that such a higher taxon-concept should change in rank at successive times, reflecting the changing status of the fundamental taxa involved and the progress of biologic evolution. In the Cretaceous to Recent angiosperm evolution (see table 5.1) the expansion from nothing to 300000 species in about 100 million years should be approximately paralleled by an expansion of the number and rank of higher taxa required. Thus in one concept a late Cretaceous species might lead to a Palaeogene genus, a Neogene family and a Recent order.

This suggests in turn that the probably very small number of Albian angiosperm 'species' should not be grouped in any taxon of such status as Magnoliopsida; it would be more appropriate instead to call them by some such name as the (family) 'Laminarphylls' in distinction from the considerable variety of 'gymnospermous' leaves among which they formed in life a very minor novel element. It

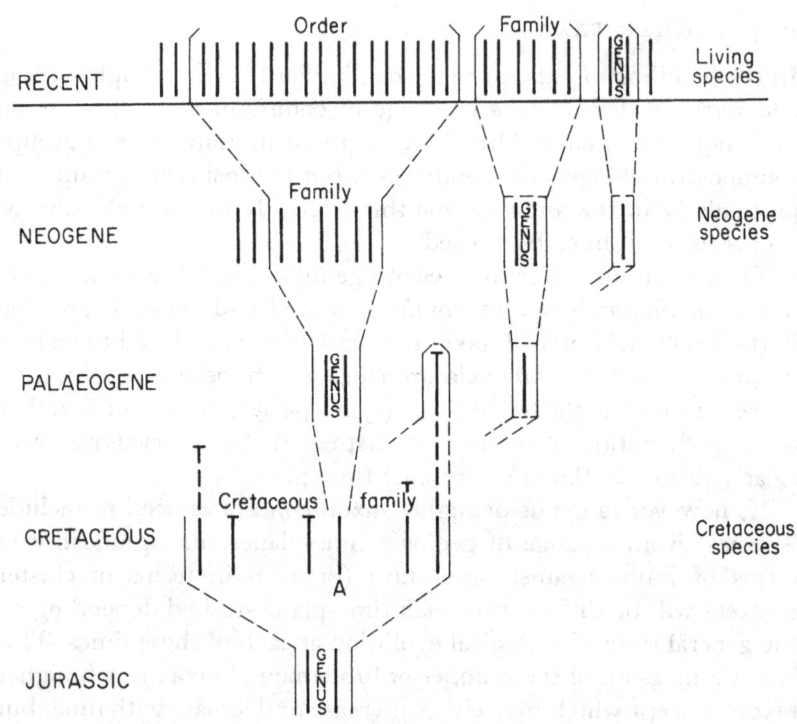

Figure 16.1. Diagram to show how the use of grades of a higher taxon concept should be time-linked and should be separately discussed and defined in each selected time-period. To keep the diagram simple it is drawn for an evolutionary radiation situation; the same principle can however apply to a reduction to extinction situation in which there would be no increase of grade with time. A single species (*A*) of Cretaceous seed-plants, although grouped as part of a Cretaceous spermatophyte family, may also be the beginning of a specialised angiosperm taxon concept leading to a living family or order. The diagram is two-dimensional for convenience.

might then also become apparent that more than one minor group name was required even in mid-Cretaceous time, and that the higher taxa of that time should probably consist of seed-plants without any reference to gymnosperms, angiosperms, Magnoliopsida etc.

Although the term 'monocotyledon' is used by Cronquist (1974) and by Doyle (1973) and although the latter use is embedded in a thought-provoking paper about Cretaceous and later fossils, the term would be better completely avoided for any Cretaceous fossil group; it seemed tempting to explain the position by use of such a name as 'dubiocot' instead of 'dicot' or 'monocot', but much better still is not to use any such term in the Cretaceous.

16.4 NEOGENE TAXONOMY

Current taxonomy and nomenclature procedures in the Neogene mostly result in the placing of isolated organ-taxa in extant groups although on the basis of very restricted character knowledge. In spite of prevailing practices in Quaternary palaeobotany, there appears to be no justification, because of the great disparity in information available, for placing any fossil whatever into an extant species. For use of an extant genus in Neogene time the justification should be in the form proposed by Hughes (1963*b*): the fossil species should be explicitly incorporated with its time connotation into the formal description of the extant genus so that the fossil cannot be ignored by workers on the various Recent species, and a continuity of fossils from the time concerned to the present day should be demonstrated.

Such a restriction may reduce the use of names in palaeoecological interpretation but this should itself force the application of evolutionary rather than uniformitarian palaeoecology.

16.5 CONCLUSION

Although discovery and interpretations of fossils will provide the principal means of establishing the course of angiosperm evolution, changes in the hierarchical side of taxonomy of the kind advocated would almost certainly strengthen and focus the search for fossils. Angiosperm origin will also be elucidated by similar treatment of early Cretaceous gymnosperms.

New knowledge of palaeolatitudes from palaeomagnetic information concerning crustal plate movements will make integrated past climate studies possible and with them some understanding of migration which was not previously possible while it was feared as a likely obstacle to time-correlation.

The variation of taxonomic treament with time which has been suggested should be kept open enough to admit the insertion of refinements as and when it becomes possible to divide the geologic time-scale in more detail.

17 Summary and future

One clear conclusion received from attempting to solve this long-standing problem is that at the present time the obstacles to understanding of the origin and early history of the angiosperms have been not so much in the intractable problems of handling fossils but mainly in the realms of human prejudice, of rigidity of working procedures, and of lack of contact between two adjacent scientific disciplines. There has also been a failure to distinguish between the necessities of palaeoecological approximation and of evolutionary discrimination. If these problems of observation can be cleared, perhaps in some of the ways suggested above, it will be possible to make contact fully again with the very real difficulties of the interpretation of the fossil evidence.

Unfortunately, following Cornford (1908), the only argument for actually doing something is that it is 'right'. In contrast there are and have been so many arguments used for doing nothing to surmount these obstacles: 'The time is not ripe to pause in work and to agree to change of methods; the present working procedures have proved sound over several generations; to strive for too great perfection is uneconomic or even unnecessary; the work of the past must be respected and not wasted; unnecessary perfection leads into a dangerous dream-land.'

Assuming that the reader has set aside all these fears, the way ahead into the fossil evidence is both exciting and challenging:

1. Joint interpretation. The growth of the integration of plant and animal information from the hitherto isolated specialities of study mainly confined to single groups of organisms.
2. Parataxa. Establishment of association formulae and confidence limits for the possible relationships of dispersed taxa.
3. Description. Techniques for continuous refinement of descriptive data.
4. Megafossils. Re-description of much Cretaceous and other early

material without the necessity to determine affinity and thus to name.

5. Microfossils. Techniques for achieving more appropriate and effective use of observation time.

6. Stratigraphy. Refinements in the art of time-correlation by means of fossils.

7. Evolution. Development of separate parallel data-handling code for palaeontology to decrease reliance on the inappropriate International Code of Botanical Nomenclature.

8. Biologic interpretation. Use of ideas from modern biological studies of living plants to broaden long-term interpretative work on fossils:

Palynology. Relationship of tetrad arrangement, pollen wall (tectum) and aperture design to pollen-tube in the pistil and to the nature and purpose of exudates (cf. Heslop-Harrison 1971). Insect and other animal flower pollenation (Faegri and Van der Pijl 1966, Proctor and Yeo 1973).

Phytochemistry. The possibility, using modern methods of micro-analysis on fossils, of detecting critical occurrences of various organic molecules in favourable preservations of dispersed plant organs.

Genetics. Relationship of degree of heterozygosity and of outcrossing potential and seed dispersal, to natural selection rate (see Proctor and Yeo 1973, p. 380).

What seldom seems to have been appreciated by geologists and palaeobiologists is that the work of the immediate future, and particularly the methods which determine the economics of that work, are immensely more important items for thought and planning than the handling of past published work which has already made at least some contribution to the current synthesis. The solution will be gradual rather than dramatic; the responsibility for pursuing it lies with those who are sufficiently trained to see the whole field clearly at the present time.

Glossary

abaxial (*Bot.*)	Of the surface of a leaf or other plant organ facing away from the adjacent stem.
Abietaceae (*Bot.*)	Conifer family including *Abies* (fir) and the pine (al. Pinaceae).
abscission (*Bot.*)	Descriptive of the breaking away of a plant organ (e.g. leaf) shed from the plant.
actinomorphic (*Bot.*)	Of a regular flower with all petals of the same shape.
adaxial (*Bot.*)	Of the surface of a leaf or other plant organ facing towards the adjacent stem.
advanced (*Biol.*)	Of a plant character believed to have arisen late in evolution.
advancement index (*Bot.*)	Compiled from the proportion of advanced characters exhibited by a family (or other taxon).
age/stage (*Geol.*)	The first level of subdivision of a geological period in the Stratigraphic Time-Scale. The term 'stage' is superfluous but has been widely used.
aggradational (*Geol.*)	Land area composed of newly deposited sediment, as in a delta.
Albian (*Geol.*)	Cretaceous age/stage time-division; see chapter 7.
amb (*Bot.*)	The outline shape of a spore or pollen grain when viewed down its polar axis; approximately = equatorial shape.
Amentiferae (*Bot.*)	Group of angiosperm families with apetalous flowers (e.g. willow, poplar, birch, walnut).
amphistomatic (*Bot.*)	Of a leaf with stomata on both faces.
anastomosing (*Bot.*)	Of the veins of a leaf which fork and re-join successively (e.g. *Sagenopteris*, fig. 8.11).
angiosperm (*Bot.*)	Seed-plant in which the ovule is completely enclosed in a carpel; for other characters see chapter 10.
aplanospore (*Bot.*)	Spores of certain green algae, which have been mistaken for pollen.
Aptian (*Geol.*)	Cretaceous age/stage time-division; see chapter 7.
aragonite (*Geol.*)	Mineral form of calcium carbonate with orthorhombic crystal symmetry.
Araucariaceae (*Bot.*)	Family of conifers with extant plants in southern hemisphere.
arborescent (*Bot.*)	Of a long-lived plant with tree form.
back-swamp (*Geol.*)	Swamp area between distributary streams behind a delta-front.

Glossary

Barremian (*Geol.*) Cretaceous age/stage time-division; see chapter 7.

belemnite (*Palaeont.*) Extinct marine cephalopod mollusc related to squid and cuttle-fish.

Benettitales (*Palaeont.*) Extinct Mesozoic gymnospermous land plants which bore flowers.

Berriasian (*Geol.*) Cretaceous age/stage time-division; see chapter 7.

binominal nomenclature (*Biol.*) Formal latinised name of plant or animal fundamental taxon consisting of generic and specific epithets.

biorecord (*Palaeont.*) Palaeontologic formal fundamental taxon (see Hughes and Moody-Stuart 1969).

bisaccate (*Bot.*) Of pollen with two separate air sacs developed within the pollen wall.

boundary point (*Geol.*) Point of definition in rock between two time-scale divisions.

Brachyphyll (*Palaeont.*) Extinct gymnosperm group with short leaves closely adpressed to the stem.

capsule (*Bot.*) Plant organ enclosing ovules.

carnivore (*Biol.*) Animal which feeds normally on other animals.

carpel (*Bot.*) Closed ovule-containing organ in the angiosperms.

cell lumen (*Bot.*) Space within dead plant cell-wall, originally occupied by the living cell.

Cenomanian (*Geol.*) Cretaceous time-division (age/stage); see chapter 7.

Ceratopsia (*Palaeont.*) Extinct Cretaceous group of horned dinosaurs.

character (*Biol.*) Recognised descriptive feature of an organism.

chemotaxonomy (*Biol.*) Classification subdivision of groups of organisms on chemical or biochemical grounds.

chert (*Geol.*) Silica accumulated secondarily in rocks, frequently enclosing and petrifying fossils.

clavae (*Bot.*) Round-headed sculpture processes on a spore or pollen exine.

coal-ball (*Geol.*) Secondary accumulation of calcium carbonate in nodules in a coal seam, often enclosing excellent petrified fossils.

colpus(i) (*Bot.*) Germinal furrow(s) on pollen grain.

comparison record (*Palaeont.*) Formal method of recording fossil data when using biorecords.

compression (*Palaeont.*) Fossil preservation in which the organism is flattened during rock consolidation.

cone-scale (*Bot.*) An independent plant organ borne on a cone-axis; a number compose a cone.

conifer (*Bot.*) Gymnosperm group in which the reproductive structures are cones.

convergent (*Biol.*) Of evolution when two or more organisms of supposed different ancestry resemble each other.

Cupressaceae (*Bot.*) Family of extant gymnosperms (e.g. cypress).

cuticle (*Bot.*) Inert water-excluding substance covering most land-plant exposed surfaces.

cutin (*Bot.*) Group of chemical materials found in cuticles.

Cycadeoidales (*Palaeont.*) Subdivision of extinct seed-plant group Benettitales.

cycadophyte (*Bot.*) Cycad-like living or fossil plant.

deciduous (*Bot.*) Of a plant in which the leaves are shed after seasons of growth to maturity.

degree of association (*Palaeont.*) Of the occurrence association of separate fossil plant organs.

denticulate (*Bot.*) Of a leaf margin with a toothed shape.

diagenesis (*Geol.*) The alterations undergone by sediments during their consolidation under gravity to form rocks.

dichotomous venation (*Bot.*) Leaf veins which show symmetrical equal forking.

dicot(yledon) (*Bot.*) Plant in which there are two seed-leaves (cotyledons) on germination of the seed.

dinoflagellates (*Bot.*) Group of motile aquatic unicellular algae in the size-range 10–200 μm.

distal pore (*Bot.*) Pore at distal pole of spore or pollen grain.

double fertilisation (*Bot.*) In angiosperms there is a separate fertilisation of a vegetative nucleus in addition to the normal activity of the pollen and egg nuclei.

endocarp (*Bot.*) Inner layer of carpel wall.

epeirogenic (*Geol.*) Vertical movements of crustal sectors without complex distortion.

ephemeral (*Bot.*) Of organisms, or organs, designed for short existence.

epigyny (*Bot.*) Character of a flower in which the ovule-bearing structure appears to lie below the cycle of petals on the axis.

epiphyte (*Bot.*) Plant which grows on another plant rather than on the ground, but is not in organic contact with the 'host'.

epistomatic (*Bot.*) Of a leaf with stomata on the upper (adaxial) surface only.

eustatic (*Geol.*) Universal changes of sea-level caused by formation or melting of ice-caps, and perhaps in other ways.

extant (*Biol.*) Of organisms living at the present day or in 'Recent' time.

exudate (*Bot.*) A fluid temporarily produced on the surface of a plant organ.

false trunk (*Bot.*) A tree trunk produced from accretion of other plant organs (usually roots) around a relatively slender stem.

'float' (*Geol.*) Geological specimen collected from scree or a slope below an outcrop.

flower (*Bot.*) Assembly of reproductive structures on a plant together with an attractant (usually visual, i.e. petals).

form-genus (*Palaeont.*) A group taxon of fossil species in which typification is for general convenience not used.

fruit (*Bot.*) Organ of a plant enclosing seeds and usually edible or otherwise transportable.

gametophyte (*Bot.*) The haploid (*n*-chromosome) part of a plant life-cycle, which may be independent in some lower plants.

genusbox (*Palaeont.*) Experimental formal taxon of generic rank which is defined by boundaries and not typified (see Hughes 1970).

geological event (*Geol.*) — A concept related to a (small) part of geological time and based on any observed or deduced geological phenomena.
ginkgophyte (*Bot.*) — Fossil plant with more resemblance to the living *Ginkgo biloba* than to other plants, living or fossil.
gonophyll (*Bot.*) — See Reference: R. Melville 1962.
gymnosperm (*Bot.*) — Seed-plant in which the ovule is at least technically exposed to the exterior by means of a micropylar tube during part of its existence, so that pollen may approach.

haplocheilic stomata (*Bot.*) — Stomata surrounded by a simple ring of subsidiary cells.
haptotypic (*Bot.*) — Of characters of the proximal face (inside in the tetrad) of a spore or pollen grain.
haustorial (*Bot.*) — Of an intrusive organ of plant cells with a function of absorbing food from the 'host'.
Hauterivian (*Geol.*) — Cretaceous age/stage time-division; see chapter 7.
Hemimetabola (*Biol.*) — Main division of insects which pass in the life-cycle through a slight metamorphosis without a pupal stage.
herb (*Bot.*) — Land plant without woody structures.
herbivore (*Biol.*) — Animal feeding exclusively on plant material.
heterozygosity (*Bot.*) — Development of a zygote or individual carrying two different alleles of a gene.
Holometabola (*Biol.*) — Main division of insects which pass in the life-cycle through a complex metamorphosis including a pupal stage.

imago (*Biol.*) — Adult (reproductive) phase of insect.
immersion (*Geol.*) — Negative epeirogenic movement of crustal sector resulting in flooding by the sea.
impregnation petrifaction (*Geol.*) — Fossil petrifaction in which cell lumina or other spaces are filled with mineral but the cell walls or existing shell crystals remain.
impression fossil (*Geol.*) — Fossil in which all organic material has been removed from a compression, leaving only a print on the rock.
inaperturate (*Bot.*) — Of a pollen grain with no detectable aperture.
inflorescence (*Bot.*) — Group of flowers on a plant.
infra-reticulum of the sacci (*Bot.*) — Structure seen in saccate pollen which is within the exine sacs.
integument (*Bot.*) — Vascularised envelope enclosing an ovule.

Kimmeridgian (*Geol.*) — Late Jurassic age/stage time-division; see chapter 7.

larval stage (*Biol.*) — Early non-reproductive stage in insect life.
Lepidoptera (*Biol.*) — Group of insects: butterflies and moths.
lianes (*Bot.*) — Woody plants of rain forests which scramble on trees but do not form free-standing trunks.
Liliatae (*Bot.*) — Formal name for the monocotyledonous group of extant angiosperms.
Linearphyll (*Palaeont.*) — Extinct gymnosperm conifer group with relatively elongated small leaves closely adpressed to the stem.

Glossary

Ma (*Geol.*)	Abbreviation for 'million years'.
maceral (*Geol.*)	Organic component of coal as a rock, comparable with a mineral but not having precise composition.
Magnoliales (*Bot.*)	Formal order of living angiosperms believed to be closely related to the *Magnolia* family.
Magnoliatae (*Bot.*)	Formal name for the whole dicotyledonous group of living angiosperms.
mangrove (*Bot.*)	Plant habitat on present margins of aggradational land close to the sea.
manoxylic (*Bot.*)	Of plant-stem wood showing a relatively loosely knit and irregular structure with wide wood rays.
megafossil (*Palaeont.*)	A fossil which is large enough to be examined without a microscope.
megaspores (*Bot.*)	Spores which on germination produce a female gametophyte; usually but not always larger than functional microspores.
megastrobilus (*Bot.*)	A cone structure (strobilus) bearing only megaspores.
Mesophytic (*Palaeont.*)	A little-used term paralleling Mesozoic but referring to the time from end-Carboniferous to mid-Cretaceous, and based on the occurrence of floras.
microflora (*Bot.*)	Flora of an environment in which the plants are microscopic. Has been used wrongly for an assemblage of microfossils.
micropyle (*Bot.*)	Tubular access through the integuments to an ovule.
microsporangiate cone (*Bot.*)	Cone bearing only microsporangia and thus only microspores or pollen.
microsporophyll (*Bot.*)	Unit of a strobilus bearing only microspores.
miospore (*Palaeont.*)	Small dispersed fossil spores or pollen of unknown function.
monocot(yledon) (*Bot.*)	Plant in which there is only one seed-leaf (cotyledon) on first germination of the seed.
monolete (*Bot.*)	Of a spore with one proximal slit-like scar.
monophyletic (*Biol.*)	Of a group of organisms believed to have descended in evolution from one immediately ancestral taxon.
monosulcate (*Bot.*)	Pollen with one linear distal germinal aperture.
mudstone (*Geol.*)	Consolidated rock of fine grain-size, but without the fissility of a shale.
multicolpate (*Bot.*)	Pollen bearing several linear apertures (colpi).
multituberculate (*Biol.*)	Descriptive of Mesozoic group of small mammals, the first herbivores of this class.
Myriapoda (*Biol.*)	Group of arthropods: millipedes and centipedes.
Neogene (*Geol.*)	Late Tertiary geological period comprising the Miocene and Pliocene.
nodal anatomy (*Bot.*)	Anatomy of that part of a stem adjacent to a leaf-base or petiole.
nucellus (*Bot.*)	Main non-reproductive tissue of the ovule before fertilisation.

oblate (*Bot.*) Pollen shape in which the equatorial diameter exceeds the length of the polar axis.

organ-genus (*Palaeont.*) Taxon of fossils of a single plant organ (e.g. leaf, pollen).

ornithopods (*Biol.*) Group of extinct Mesozoic dinosaurs with bipedal gait.

orogeny (*Geol.*) Disturbance of crustal rock stratification involving lateral pressure and distortion as well as uplift.

outcrop (*Geol.*) Part of rock body visible at surface of the earth.

ovule (*Bot.*) Protective structure enclosing the plant megaspore, which develops into a seed.

Palaeogene (*Geol.*) Early Tertiary geological period comprising Palaeocene, Eocene, Oligocene.

palaeolatitude (*Geol.*) Latitude determined for a specified past time.

palaeopalynology (*Palaeont.*) Palynology of fossils.

palmate (*Bot.*) Of leaves composed of radially arranged segments separately borne on the petiole.

palynomorph (*Palaeont.*) Term used for fossils resembling spores and pollen, but not so identified.

pappus (*Bot.*) Tuft of hairs borne on a seed.

parataxa (*Palaeont.*) Taxa of fossils used for remains of only part of the organism or part of the life-cycle.

pauci-aperturate (*Bot.*) Angiosperm pollen with three or less apertures.

peltate (*Bot.*) Of a plant organ with a swollen head, e.g. certain cone-scales.

perianth (*Bot.*) That part of a flower below and outside the stamens (which carry the anthers).

period/system (*Geol.*) Large geological time-scale division/rocks laid down in that division of the time-scale. The concepts are interdependent.

permeable (*Geol.*) Rock through which fluid under a hydrostatic head will pass.

petiole (*Bot.*) Narrow strengthened basal part of a leaf (leaf stalk).

phylogeny (*Biol.*) A postulated evolutionary sequence (in time) of descent from one taxon to another.

Pinaceae (*Bot.*) Family of living conifers, e.g. pine, fir. (Syn. Abietaceae.)

pinnate (*Bot.*) Of a leaf which is compound with small leaflets (pinnules) arranged symmetrically on a rachis.

pistil (*Bot.*) Central organ of the flower, composed of one or more carpels.

plate tectonics (*Geol.*) Explanation of lateral crustal movements involving the whole lithosphere which includes part of the upper mantle.

pollen chamber (*Bot.*) Space enclosed by integument just above the megaspore end of the ovule, and beneath the micropyle.

pollen-tube (*Bot.*) On germination a pollen-tube breaks out of the exine and through this the reproductive nuclei are transmitted.

polyphyletic (*Biol.*) Group of fossils or organisms believed not to be homogeneous in terms of single direct evolutionary origin.

polyploid (*Bot.*) Plants which carry three or more sets of homologous chromosomes in the adult plant (sporophyte) phase.

polyporate (*Bot.*)	Angiosperm pollen with numerous pore-like apertures.
porate (*Bot.*)	Pollen with equidimensional aperture or apertures.
Portland beds (*Geol.*)	Late Jurassic strata in southern England, typically in Dorset.
primitive (*Biol.*)	Of a character of an organism believed to have arisen early in evolution.
priority (*Biol.*)	In nomenclature priority is given to the first description of a taxon and is maintained against other descriptions.
protandry (*Bot.*)	In a flower, the development to maturity of the male organs before the ovules.
pteridophyte (*Bot.*)	A group of spore-bearing land plants including the ferns.
pteridosperm (*Bot.*)	A group of extinct seed-plants with fern-like leaves.
pulmonate (*Biol.*)	A type of land gastropod (snail) with a 'lung' for gaseous exchange.
pupal stage (*Biol.*)	Insect resting stage in life-cycle between larva and imago; in the Holometabola only.
Purbeck beds (*Geol.*)	Latest Jurassic and earliest Cretaceous strata in southern England, typically in Dorset.
pycnoxylic (*Bot.*)	Of wood which is dense and composed of regularly arranged tracheids with narrow wood rays.
pyrite (*Geol.*)	A common form of iron sulphide which forms petrifactions.
pyroclastic (*Geol.*)	A sediment of volcanic debris.
rachis (*Bot.*)	Extension of the petiole to which leaflets are attached in a pinnately compound leaf.
Ranales (*Bot.*)	Group of angiosperm families including Magnoliaceae, Ranunculaceae etc.
relict taxon (*Biol.*)	Taxon of organisms isolated on a time-plane and believed to be a survivor in evolution from a larger group of taxa.
reticulate (*Bot.*)	Of leaf venation when a rectangular network is formed.
retipilate (*Bot.*)	Of a reticulum-like pattern composed of pila (hairs) on the exine of certain angiosperm pollen.
reworking (*Geol.*)	Of fossils, is their natural removal from one sediment and incorporation in another.
rimula (*Palaeont.*)	Distal ring-furrow on the fossil pollen grain *Classopollis.*
rockleach (*Geol.*)	Solution and downward transport under gravity of fossil and mineral material by percolating water in the unsaturated zone above the water-table.
rock unit (*Geol.*)	Rock mass defined and delimited for mapping purposes.
saccate pollen (*Bot.*)	Pollen with one or more air sacs developed in the exine.
sapromyophily (*Bot.*)	Pollination mechanism of the flower in which flies are attracted by deceit.
sauropods (*Biol.*)	Group of giant Mesozoic semi-aquatic dinosaurs.
scalariform (*Bot.*)	Of transversely elongated pits in the wall of a xylem vessel, reminiscent of a ladder in appearance.
sculpture (*Bot.*)	Of pollen exine (or other organs), the elements raised above the general surface.
seed (*Bot.*)	The enduring reproductive structure of a plant, capable of long dormancy.

siphonogamy (*Bot.*) Seed-plant transfer of male gametes by means of a pollen tube.
sporopollenin (*Bot.*) Material of pollen and spore exines, chemically oxidative polymers of carotenoids.
stigma (*Bot.*) The part of the flower that receives the pollen.
stoma(ta) (*Bot.*) Opening through plant cuticle for gaseous exchange, controlled by two adjacent guard cells.
structure (*Bot.*) Of pollen exine, detailed features within the pollen wall.
subsidiary cells (*Bot.*) Adjacent epidermal cells immediately surrounding a stoma and its guard cells.
sulcus (*Bot.*) Single germinal furrow confined to distal face of pollen grain.
syndetocheile stoma (*Bot.*) Special type of stoma in which only two subsidiary cells flank the two guard cells, e.g. in Benettitales.
synonyms (*Biol.*) Names given to a taxon which are no longer valid under the Code of Nomenclature.
systematics (*Biol.*) The process of classifying and naming organisms.

Taxaceae (*Bot.*) Family of extant gymnosperms resembling conifers but distinct in their reproductive structures (e.g. yew tree).
taxon (pl. taxa) (*Biol.*) A classificatory unit.
taxonomy (*Biol.*) The process of dividing up organisms into taxa.
tectum (*Bot.*) Outer covering layer of pollen exine in many angiosperms.
tetrad (*Bot.*) Group of four spores or pollen grains resulting from two successive regular divisions of the spore mother-cell.
time-correlation (*Geol.*) Correlation in time-sequence between events raised from two rock columns.
time-scale (*Geol.*) A stratigraphic sequence time-scale is set up for reference, and exists independently of its calibration.
Tithonian (*Geol.*) Late Jurassic age/stage time-division; see chapter 7.
transgression (*Geol.*) Inundation of land by sea, normally an immersion of epeirogenic origin.
tricolpate (*Bot.*) Angiosperm pollen with three elongate colpi (apertures) radially arranged.
tricolpodiorate (*Bot.*) Angiosperm tricolpate pollen with two pores per colpus.
tricolporate (*Bot.*) Angiosperm tricolpate pollen with a pore developed equatorially in each colpus.
tricolporoid (*Bot.*) Angiosperm tricolpate pollen in incipient tricolporate state.
Turonian (*Geol.*) Late Cretaceous age/stage time-division; see chapter 7.
typification (*Biol.*) In Linnean taxonomy, each taxon of family and lower status is referred in nomenclature to a holotype, type species or type genus.

unconformity (*Geol.*) Discordance between two sets of layered rocks produced by a period of uplift and erosion after the deposition of the first set.
uniformitarianism (*Geol.*) Hypothesis (erroneous) that past processes were often identical with present processes and so can be interpreted.

unilocular (*Bot.*) Of an angiosperm ovary with only one cavity (locule) containing ovules.

upland (*Geol.*) Land surface suffering net erosion.

Valanginian (*Geol.*) Cretaceous age/stage time-division; see chapter 7.

vessel (*Bot.*) A line of lignified water-conducting cells made continuous by absorption of the end-walls.

Volgian (*Geol.*) Late Jurassic age/stage time-division in northern hemisphere.

water-table (*Geol.*) Natural standing level of underground water is continuous with the sea and lake surfaces.

Wealden (*Geol.*) Group of early Cretaceous non-marine rocks outcropping in southern England.

wood-pitting (*Bot.*) Thin areas in the walls of the wood cells of a stem for interconnection between adjacent cells.

wood ray (*Bot.*) Vertical radial sheets of conducting tissue running through a woody stem.

xeromorphic (*Bot.*) A plant structure apparently suited to survival of the plant in drought.

zonosulcate (*Bot.*) Of pollen, with a ring sulcus.

zooidogamy (*Biol.*) Release of free gametes into the pollen chamber prior to fertilisation, as in living cycads.

zygomorphic (*Bot.*) Of a flower with only bilateral symmetry or with none.

References

Allen, P., Keith, M. L., Tan, F. C. and Deines, P. (1973). Isotopic ratios and Wealden environments. *Palaeontology*, **16**, 607–22.

Alvin, K. L. (1953). Three Abietaceous cones from the Wealden of Belgium. *Mém. Inst. roy. Sci. Nat. Belg.* **125**, 42 pp., 5 pl.

Alvin, K. L. (1960*a*). On the seed *Vesquia tournaisii* C. E. Bertrand, from the Belgian Wealden. *Ann. Bot.* (N.S.) **24**, 508–15.

Alvin, K. L. (1960*b*). Further conifers of the Pinaceae from the Wealden formation of Belgium. *Bull. Inst. Roy. Sci. Nat. Belgique*, **146**, 5–38.

Alvin, K. L. and Chaloner, W. G. (1970). Parallel evolution in leaf venation: an alternative view of angiosperm origins. *Nature*, **226**, 662–3.

Alvin, K. L., Barnard, P. D. W., Harns, T. M., Hughes, N. F., Wagner, R. H. and Wesley, A. (1967). Gymnospermophyta. In Harland, W. B. *et al.* (eds.) *The Fossil Record*, Geological Society, London, pp. 247–68.

Anderson, F. W. and Hughes, N. F. (1964). The 'Wealden' of North-West Germany and its English equivalents. *Nature*, **201**, 907–8.

Archangelsky, S. A. (1965). Fossil Ginkgoales from the Tico flora, Santa Cruz Province, Argentina. *Bull. Brit. Mus.* (*Nat. Hist.*) *Geol.* **10**, 121–37, 5 pl.

Archangelsky, S. A. (1968). On the genus Tomaxiella (Coniferae). *J. Linn. Soc. Lond. Bot.* **61**, 153–65, 4 pl.

Archangelsky, S. A. and Gamerro, J. C. (1967). Pollen grains found in coniferous cones from the Lower Cretaceous of Patagonia (Argentina). *Rev. Palaeobotan. Palynol.* **5**, 179–82.

Axelrod, D. I. (1959). Poleward migration of early angiosperm floras. *Science*, **130**, 203–7.

Axelrod, D. I. (1967). Drought, diastrophism, and quantum evolution. *Evolution*, **21**, 201–9.

Axelrod, D. I. (1972). Edaphic aridity as a factor in angiosperm evolution. *American Naturalist*, **106**, 311–20.

Bannan, M. W. and Fry, W. L. (1957). Three Cretaceous woods from the Canadian Arctic. *Can. J. Bot.* **35**, 327–37, 4 pl.

Barghoorn, E. S. (1951). Age and environment: a survey of North American Tertiary floras in relation to palaeoecology. *J. Palaeont.* **25**, 736–44.

Barnard, P. D. W. (1968). A new species of *Masculostrobus* Seward producing *Classopollis* pollen, from the Jurassic of Iran. *J. Linn. Soc. Lond. Bot.* **61**, 167–76, 1 pl.

Bate-Smith, E. C. (1968). The phenolic constituents of plants and their taxonomic significance, 2. Monocotyledons. *J. Linn. Soc. Lond. Bot.* **60**, 325–56.

Batten, D. J. (1973). Use of palynologic asemblage-types in Wealden correlation. *Palaeontology*, **16**, 1–40, 2 pl.

References

Becker, H. F. (1971). New plant structures from the Triassic Dolores Formation in south western Colorado. *Amer. J. Bot.* **58**, 467–8 (abstract).

Bell, W. A. (1956). Lower Cretaceous floras of Western Canada. *Mem. geol. Surv. Can.* **285**, 331 pp.

Belsky, C. Y., Boltenhagen, E. and Potonié, R. (1965). Sporae dispersae der Oberen Kreide von Gabon, Äquatoriales Afrika. *Paläont. Z.* **39**, 72–83, 2 pl.

Berry, E. W. (1911). Systematic paleontology of the Lower Cretaceous deposits of Maryland. In Clark, W. B. *et al. Maryland Geol. Surv.: Lower Cretaceous*, 179–596.

Berry, E. W. (1916). The Upper Cretaceous floras of the world. *Maryland Geol. Surv.* 183–314.

Berry, E. W. (1922). The flora of the Cheyenne Sandstone of Kansas. *US Geol. Surv. Prof. Paper*, **127**, 199–225.

Binfield, W. R. and Binfield, H. (1854). On the occurrence of fossil insects in the Wealden strata of the Sussex coast. *Q. Jl geol. Soc. Lond.* **10**, 171–6.

Black, M. (1929). Drifted plant beds of the Upper Estuarine Series of Yorkshire. *Q. Jl geol. Soc. Lond.* **85**, 389–439.

Bolkhovitina, N. A. (1959). Spore-pollen complexes of the Mesozoic deposits in the Vilué Basin and their stratigraphic significance. *Tr. geol. Inst. Leningrad* **24**, 1–185. (In Russian.)

Boltenhagen, E. (1963). Etat des connaissances en palynologie du Crétacé inférieur. *Mém. Bur. Rech. Géol. Minières*, **34**, 555–75.

Boltenhagen, E. (1967). Spores et pollen du Crétacé Supérieur du Gabon. *Pollen et Spores*, **9**, 335–55.

Bommer, Ch. (1891). Sur le gîte wealdien à végétaux de Braquegnies (Hainaut). *Bull. Soc. belge Geol. Paléont. Hydr.* **5**, 196–7.

Bommer, Ch. (1892). Un nouveau gîte de végétaux découvert dans l'argile Wealdienne de Braquegnies (Hainaut). *Bull. Soc. belge Géol. Paléont. Hydr.* **6**, 160–61.

Boureau, E. (1954). Sur le *Palmoidopteris lapportii* n.g., n.sp. *Mém. Mus. Nat. Hist.* N.S. 3, fasc. 3.

Brenner, G. J. (1963). The spores and pollen of the Potomac Group of Maryland. *Bull. Md. Dept. Geol. Mines*, **27**, 1–215.

Brenner, G. J. (1967*a*). Early angiosperm pollen differentiation in the Albian to Cenomanian deposits of Delaware (USA). *Rev. Palaeobotan. Palynol.* **1**, 219–27, 3 pl.

Brenner, G. J. (1967*b*). The gymnospermous affinity of *Eucommiidites* Erdtman 1948. *Rev. Palaeobotan. Palynol.* **5**, 123–7, 1 pl.

Brenner, G. J. (1968). Middle Cretaceous spores and pollen from north-eastern Peru. *Pollen et Spores*, **10**, 341–54.

Brodie, P. B. (1854). On the insect-beds of the Purbeck formation in Wiltshire and Dorsetshire. *Q. Jl geol. Soc. Lond.* **10**, 475–82.

Brown, R. W. (1956). Palm-like plants from the Dolores formation (Triassic), South-western Colorado. *US Geol. Surv. Prof. Paper*, 274-H, 205–9.

Burakova, A. T. (1971). *Sogdiania abdita* from the Middle Jurassic of Central Asia – a probable ancestor of flowering plants. *Voprosy Paleontologii* (*Leningrad*), **4**, 3–7.

Burger, D. (1966). Palynology of Uppermost Jurassic and Lowermost Cretaceous strata in the Eastern Netherlands. *Leidse Geol. Med.* **35**, 209–76.

Casey, R. (1961). The stratigraphical palaeontology of the Lower Greensand. *Palaeontology*, **3**, 487–621, 8 pl.

Casey, R. (1964). The Cretaceous period. In Harland, W. B. *et al.* (eds.) The Phanerozoic Time-scale. *Q. Jl geol. Soc. Lond.* **120 S**, 193–202.

Chaloner, W. G. (1958). The Carboniferous upland flora. *Geol. Mag.* **95**, 261–2.

Chaloner, W. G. (1970). The evolution of miospore polarity. *Geoscience and Man*, **1**, 47–56, 1 pl.

Chaloner, W. G. and Creber, G. T. (1973). In Tarling, D. H. and Runcorn, S. K. (eds.) *Implications of Continental Drift to the Earth Sciences*. Academic Press, New York.

Chaloner, W. G. and Muir, M. (1968). Spores and Floras. In Murchison, D. G. and Westoll, T. S. (eds.) *Coal and Coal-bearing Strata*, Oliver and Boyd, Edinburgh, pp. 127–46.

Chandler, M. E. J. (1954). Upper Cretaceous *Icacinocarya* fruits from Egypt. *Bull. Brit. Mus.* (*Nat. Hist.*) *Geol.* **2**, 149–87.

Chandler, M. E. J. (1958). Angiosperm fruits from the Lower Cretaceous of France and the Lower Eocene of Germany. *Ann. Mag. Nat. Hist.* (Ser. 13) **1**, 354–8.

Chandler, M. E. J. (1966). Fruiting organs from the Morrison formation of Utah, USA. *Bull. Brit. Mus.* (*Nat. Hist.*) *Geol.* **12**, 139–71, 12 pl.

Chandler, M. E. J. and Axelrod, D. I. (1961). An early angiosperm fruit from California. *Am. J. Sci.* **259**, 441–6.

Chesters, K. I. M., Gnauck, F. R. and Hughes, N. F. (1967). Angiospermae. In Harland, W. B. *et al.* (eds.) *The Fossil Record*, Geological Society, London, pp. 269–88.

Clark, W. B., Bibbins, A. B. and Berry, E. W. (1911). *Maryland Geological Survey: Lower Cretaceous*, The Johns Hopkins Press, Baltimore. 622 pp., 97 pl.

Cornford, F. M. (1908). *Microcosmographia Academia*, Bowes, Cambridge.

Couper, R. A. (1956). Evidence for a possible Gymnospermous affinity for *Tricolpites troedssonii* Erdtman. *New Phytol.* **55**, 2, 280–85.

Couper, R. A. (1958). British Mesozoic microspores and pollen grains, a systematic and stratigraphic study. *Palaeontographica*, *B* **103**, 75–179.

Couper, R. A. and Hughes, N. F. (1963). Jurassic and Lower Cretaceous palynology of the Netherlands and adjacent areas. *Verh. K. ned. geol-mijnb. Genoot.* **21**, 105–8, 3 pl.

Creber, G. T. (1972). Gymnospermous wood from the Kimmeridgian of East Sutherland and from the Sandringham Sands of Norfolk. *Palaeontology*, **15**, 655–61.

Cridland, A. A. (1964). *Amyelon* in American coal balls. *Palaeontology*, **7**, 186–209.

Cronquist, A. (1968). *The Evolution and Classification of Flowering Plants.* Houghton Mifflin Co., Boston.

Cronquist, A. (1974). Thoughts on the origin of Monocotyledons. *Birbal Sahni Inst. Palaeobotan. Spec. Publ.* (Lucknow), **1**, 19–24.

Crowson, R. A. (1962). Observations on the beetle family Cupedidae. *Ann. Mag. Nat. Hist.* (Ser. 13), **5**, 147–57, 2 pl.

Crowson, R. A., Smart, J. and Wootton, R. J. (1967). Insecta. In Harland, W. B. *et al.* (eds.) *The Fossil Record*, Geological Society, London, pp. 508–28.

References

Cutbill, J. L. and Funnell, B. M. (1967). Numerical analysis of *The Fossil Record*. In Harland, W. B. *et al.* (eds.) *The Fossil Record*, Geological Society, London, pp. 791–820.

Daber, R. (1968). A *Weichselia–Stiehleria*–Matoniaceae Community within the Quedlinburg estuary of Lower Cretaceous age. *J. Linn. Soc. Lond. Bot.* **61**, 75–85, 2 pl.

Davis, P. H. and Heywood, V. H. (1966). *Principles of Angiosperm Taxonomy*, Oliver and Boyd, Edinburgh, 556 pp.

Delevoryas, T. (1962). *Morphology and Evolution of Fossil Plants*, Holt, Rinehart and Winston, New York, 189 pp.

Delevoryas, T. (1968). Some aspects of cycadeoid evolution. *J. Linn. Soc. Lond. Bot.* **61**, 137–46.

Delevoryas, T. and Gould, R. E. (1971). An unusual fossil fructification from the Jurassic of Oaxaca, Mexico. *Amer. J. Bot.* **58**, 616–20.

Dettmann, M. E. (1963). Upper Mesozoic microfloras from south-eastern Australia. *Proc. Roy. Soc. Victoria*, **77**, 1–148.

Dettmann, M. E. (1973). Angiospermous pollen from Albian to Turonian sediments of Eastern Australia. *Spec. Publ. Geol. Soc. Austr.* **4**, 3–34, 6 pl.

Dilcher, D. L. (1974). Approaches to the identification of angiosperm leaf remains. *Bot. Rev.* (New York), **40**, 1–157.

Döring, H. (1965). Die Sporenpaläontologische Gliederung des Wealden in Westmecklenburg (Structur Werle). *Geologie Jg.* **14**, suppl. 47, 1–118, 23 pl.

Doyle, J. A. (1969). Cretaceous angiosperm pollen of the Atlantic coastal plain and its evolutionary significance. *J. Arnold Arb.* **50**, 1–35.

Doyle, J. A. (1973). Fossil evidence on early evolution of the monocotyledons. *Q. Rev. Biol.*, **48**, 399–413.

Doyle, J. A. and Hickey, L. J. (1972). Coordinated evolution in Potomac Group angiosperm pollen and leaves. *Amer. J. Bot.* **59**, 660 (abstract).

Edmunds, F. H. (1935). The Wealden District: British Regional Geology. Geol. Svy., London.

Edwards, W. N. (1921). On a small benettitalean flower from the Wealden of Sussex. *Ann. Mag. Nat. Hist.* (Ser. 9), **8**, 440–2, 1 pl.

Ehrendorfer, F., Krendl, F., Hábeler, E. and Sauer, W. (1968). Chromosome numbers and evolution in primitive angiosperms. *Taxon*, **17**, 337–53.

Erdtman, G. (1948). Did dicotyledonous plants exist in early Jurassic times? *Geol. Foren. Forhandl.* **70**, 265–71.

Erdtman, G. (1952). *Pollen Morphology and Plant Taxonomy, Angiosperms*, Almquist and Wiksell, Stockholm, 539 pp.

Evans, H. E. (1969). Three new Cretaceous aculeate wasps (Hymenoptera). *Psyche*, Cambridge, Mass., **76**, 251–61.

Eyde, R. H. (1972). Note on geologic histories of flowering plants. *Brittonia*, **24**, 111–16.

Faegri, K. and van der Pijl, L. (1966). *Principles of Pollination Ecology*, Pergamon Press, Oxford, 248 pp.

Fisher, J. M. McC. (1967). Fossil birds and their adaptive radiation. In Harland, W. B. *et al.* (eds.) *The Fossil Record*, Geological Society, London, pp. 135–54.

Fliche, P. (1905). Note sur des bois fossiles de Madagascar. *Bull. soc. géol. France*, **5**, 346 pp.

Florin, R. (1958). On Jurassic taxads and conifers from North Western Europe and Greenland. *Acta Hort. Bergian.* **17**, 257–402.

Florin, R. (1963). The distribution of conifer and taxad genera in time and space. *Acta Hort. Bergian.* **20**, 121–312.

Fontaine, W. M. (1889). The Potomac or Younger Mesozoic Flora. *Monogr. U.S. geol. Surv.* **15**, 377 pp.

Foster, A. S. and Gifford, E. M. (1959). *Comparative Morphology of Vascular Plants*, Freeman, San Francisco, 555 pp.

Frederick, J. F. (1970). Phylogeny and morphogenesis in the Algae. *Science*, **169**, 403–4.

Frederiksen, N. O. (1972). The rise of the Mesophytic flora. *Geoscience and Man*, **4**, 17–28.

Gallois, R. W. (1965). The Wealden District: British Regional Geology, 4th edition. Geol. Svy., London.

Galton, P. M. (1974). The ornithischian dinosaur *Hypsilophodon* from the Wealden of the Isle of Wight. *Bull. Brit. Mus. (Nat. Hist.) Geol.* **25**, 152 pp., 2 pl.

Gamerro, J. C. (1965). Morfologia del polen de la conifera *Trisacocladus tigrensis* Archang. de la Formación Bagneró, Provincia de Santa Cruz. *Ameghiniana*, **4**, 31–8.

Goczan, F., Groot, J. J., Krutzsch, W. and Pacltová, B. (1967). Die Gattungen des 'Stemma Normapolles Pflug 1953*b*' (Angiospermae). *Paläont. Abh.* **2B**, 429–539.

Graham, S. A. and Graham, A. (1971). Palynology and systematics of *Cuphea* (Lythraceae), 2. Pollen morphology and infrageneric classification. *Amer. J. Bot.* **58**, 844–57.

Grant, V. (1963). *The Origin of Adaptation*, Columbia University Press, New York and London.

Groot, J. J. and Penny, J. S. (1960). Plant microfossils and age of non-marine Cretaceous sediments of Maryland and Delaware. *Micropaleontology*, **6**, 225–36, 2 pl.

Groot, J. J. and Groot, C. R. (1962). Plant microfossils from Aptian, Albian and Cenomanian deposits of Portugal. *Com. Serv. Geol. Portugal*, **46**, 131–71, 10 pl.

Habib, D. (1968). Spores, pollen and microplankton from the Horizon Beta outcrop. *Science*, **162**, 1480–81.

Habib, D. (1969). Middle Cretaceous palynomorphs in a deep sea core from the Seismic Reflector Horizon *A* outcrop area. *Micropalaeontology*, **15**, 85–101, 4 pl.

Hancock, J. M. (1967). Some Cretaceous–Tertiary marine faunal changes. In Harland, W. B. *et al.* (eds.) *The Fossil Record*, Geological Society, London, pp. 91–104.

Handlirsch, A. (1906–08). *Die fossile Insekten und die Phylogenie der Rezenten Formen*, Vol. 1, Leipzig.

Harborne, J. B. (ed.) (1969). *Phytochemical Phylogeny*, Academic Press, New York.

Harland, W. B. and Hacker, J. L. F. (1966). 'Fossil' lightning strikes 250 million years ago. *Adv. Sci.* 663–71.

Harland, W. B. *et al.* (eds.) (1967). *The Fossil Record*, Geological Society, London, 828 pp.

Harland, W. B. *et al.* (1972). A concise guide to stratigraphical procedure. *Jl geol. Soc. Lond.* **128**, 295–305.

Harris, T. M. (1932). Fossil flora of Scoresby Sound, 2. *Medd. om Grønl.* **85**, 1–112.

References

Harris, T. M. (1942). *Wonnacottia*, new Bennettitalean microsporophyll. *Ann. Bot.* N.S. **6**, 577–92.

Harris, T. M. (1943). The fossil conifer *Elatides williamsonii*. *Ann. Bot.* N.S. **7**, 325–39.

Harris, T. M. (1945). On a coprolite of *Caytonia* pollen. *Ann. Mag. Nat. Hist.* (Ser. 11), **12**, 373–8.

Harris, T. M. (1948). Notes on the Jurassic Flora of Yorkshire, 37–39. *Ann. Mag. Nat. Hist.* (Ser. 12), **1**, 181–213.

Harris, T. M. (1951 *a*). The fructification of *Czekanowskia* and its allies. *Phil. Trans. roy. Soc. B* **235**, 483–508.

Harris, T. M. (1951 *b*). Notes on the Jurassic Flora of Yorkshire, 49–51. *Ann. Mag. Nat. Hist.* (Ser. 12), **4**, 915–37.

Harris, T. M. (1952). Notes on the Jurassic Flora of Yorkshire, 52–54. *Ann. Mag. Nat. Hist.* (Ser. 12), **5**, 362–82.

Harris, T. M. (1953 *a*). The geology of the Yorkshire Jurassic flora. *Proc. Yorkshire Geol. Soc.* **29**, 63–71.

Harris, T. M. (1953 *b*). Conifers of the Taxodiaceae from the Wealden formation of Belgium. *Inst. Roy. Sci. Nat. Belgique, Mém.* **126**, 43 pp., 8 pl.

Harris, T. M. (1954). Mesozoic seed cuticles. *Svensk. Botan. Tidskr.* **48**, 281–91.

Harris, T. M. (1956). The mystery of flowering plants. *Listener*, London, 514–16.

Harris, T. M. (1957). A Liasso-Rhaetic flora in South Wales. *Proc. Roy. Soc. London, B* **147**, 289–308.

Harris, T. M. (1958). Forest fire in the Mesozoic. *J. Ecol.* **46**, 447–53.

Harris, T. M. (1961). The fossil Cycads. *Palaeontology*, **4**, 313–23.

Harris, T. M. (1962). The occurrence of the fructification *Carnoconites* in New Zealand. *Trans. Roy. Soc. N.Z.* **1**, 17–27.

Harris, T. M. (1964). *The Yorkshire Jurassic Flora* **2**; *Caytoniales, Cycadales* and *Pteridosperms*. Brit. Mus. (Nat. Hist.), 191 pp., 7 pl.

Harris, T. M. (1969). *The Yorkshire Jurassic Flora* **3**; *Bennettitales*. Brit. Mus. (Nat. Hist.), 186 pp., 7 pl.

Harris, T. M. (1974). *Williamsoniella lignieri*: its pollen and the compression of spherical pollen grains. *Palaeontology*, **17**, 125–48.

Hedlund, R. W. (1966). Palynology of the Red Branch member of the Woodbine Formation (Cenomanian), Bryan County, Oklahoma. *Oklahoma Geol. Surv. Bull.* **112**, 69 pp.

Hedlund, R. W. and Norris, G. (1968). Spores and pollen grains from Fredericksburgian (Albian) strata, Marshall County, Oklahoma. *Pollen et Spores*, **10**, 129–59, 9 pl.

Heer, O. (1868). *Die fossile Flora der Polarländer*, vol. 1, Zurich, 192 pp., 50 pl.

Heer, O. (1874). Flora Fossilis Arctica, vol. 3: Nachträge zur fossilen Flora Grönlands, *Kgl. Svenska Vet. Handl.* **13**, 29 pp.

Helal, A. (1965). Jurassic spores and pollen grains from the Kharga oasis, western desert, Egypt. *Neues Jahrb. Geol. Paläontol. Abhandl.* **123**, 160–66.

Hennig, W. (1970). Insektfossilien aus der untere Kreide, 2. Empididae (Diptera, Brachycera). *Stutt. Beitr. Naturk.* **214**, 1–12.

Herngreen, G. F. W. (1973). Palynology of Albian–Cenomanian strata of borehole 1-QS-1-MA, State of Maranhao, Brazil. *Pollen et Spores*, **15**, 515–55.

Herngreen, G. F. W. (1974). Middle Cretaceous palynomorphs from Northeastern Brazil. *Sci. Géol. Bull.*, Strasbourg, **27**, 1–2, 101–16, 2 pl.

Heslop-Harrison, J. (1971). *Pollen: Development and Physiology*, Butterworths, London, 338 pp.
Heslop-Harrison, J. (1975). The adaptive significance of the exine architecture. *Linn. Soc. Sympos.*, Academic Press.
Hickey, L. J. (1971). Evolutionary significance of leaf architectural features in the woody dicots. *Amer. J. Bot.* **58**, 469 (abstract).
Hickey, L. J. and Doyle, J. (1972). Fossil evidence on evolution of angiosperm leaf venation. *Amer. J. Bot.* **59**, 661 (abstract).
Hirmer, M. and Hörhammer, L. (1934). Zur weiteren Kenntnis von *Cheirolepis* Schimper and *Hirmeriella* Hörhammer mit Bemerkungen über deren systematische Stellung. *Palaeontographica*, Stuttgart, **79***B*, 67–8, 8 pl.
Hudson, J. D. and Palframan, D. F. B. (1969). The ecology and preservation of the Oxford Clay fauna at Woodham, Buckinghamshire. *Q. Jl geol. Soc. Lond.* **124**, 387–418, 2 pl.
Hughes, N. F. (1958). Palaeontological evidence for the age of the English Wealden. *Geol. Mag.* **95**, 41–9.
Hughes, N. F. (1961*a*). Fossil evidence and angiosperm ancestry. *Sci. Progr.* **48**, 84–102.
Hughes, N. F. (1961*b*). Further interpretation of *Eucommiidites* Erdtmann 1948. *Palaeontology*, **4**, 292–9.
Hughes, N. F. (1963*a*). Cretaceous Floras and the Assessment of Past Climates. In Nairn, A. E. M. (ed.) *Problems in Palaeoclimatology*, Interscience, New York, 705 pp.
Hughes, N. F. (1963*b*). The assignment of species of fossils to genera. *Taxon*, **12**, 336–7.
Hughes, N. F. (1971). Remedy for the general data-handling failure of palaeontology. In Cutbill, J. L. (ed.) Data processing in biology and geology. *Syst. Assoc. Spec.* **3**, 321–30.
Hughes, N. F. (1973*a*). Mesozoic and Tertiary distributions and problems of land-plant evolution. In Hughes, N. F. (ed.) Organisms and continents through time. *Spec. Pap. Palaeont.* **12**, 189–98.
Hughes, N. F. (1973*b*). Environment of angiosperm origins. In Palynology of Mesophyte. *Proc. 3 Internat. Palynol. Sympos.* (Novosibirsk), Nauka, Moscow, pp. 135–7.
Hughes, N. F. (1973*c*). Towards effective data-handling in palaeopalynology. In Morphology and systematics of fossil pollen and spores. *Proc. 3 Internat. Palyn. Sympos.* (Novosibirsk), Nauka, Moscow, pp. 9–14.
Hughes, N. F. (1974). Angiosperm evolution and the superfluous upland origin hypothesis. *Birbal Sahni inst. Palaeobotan. Spec. Publ.* (Lucknow), **1**, 25–9.
Hughes, N. F. (1975). Cretaceous palaeobotanic problem. In Beck, C. B. (ed.) *Origin and Early Evolution of Angiosperms*. Columbia University Press, New York and London.
Hughes, N. F. and Couper, R. A. (1958). Palynology of the Brora Coal of the Scottish Middle Jurassic. *Nature*, **181**, 1482–3.
Hughes, N. F. and Croxton, C. A. (1973). Palynologic correlation of the Dorset 'Wealden'. *Palaeontology*, **16**, 567–601, 10 pl.
Hughes, N. F. and Moody-Stuart, J. C. (1967*a*). Palynological facies and correlation in the English Wealden. *Rev. Palaeobotan. Palynol.* **1**, 259–68.

References

Hughes, N. F. and Moody-Stuart, J. C. (1967*b*). Proposed method of recording pre-Quaternary palynological data. *Rev. Palaeobotan. Palynol.* **3**, 347–58, 1 pl.

Hughes, N. F. and Moody-Stuart, J. C. (1969). A method of stratigraphic correlation using early Cretaceous microspores. *Palaeontology*, **12**, 84–111.

Hughes, N. F. and Smart, J. (1967). Plant–insect relationships in Palaeozoic and later time. In Harland, W. B. *et al.* (eds.) *The Fossil Record*, Geological Society, London, pp. 107–17.

Jardiné, S. (1967). Spores à expansions en forme d'élatères du Crétacé moyen d'Afrique Occidentale. *Rev. Palaeobotan. Palynol.* **1**, 235–58.

Jardiné, S. and Magloire, L. (1965). Palynologie et stratigraphie du Crétacé des bassins du Sénégal et de Côte d'Ivoire. *Mem. Bur. Rech. Géol. Minières*, **32**, 187–245, 11 pl.

Joysey, K. A. (1956). The nomenclature and comparison of fossil communities. In P. C. Sylvester-Bradley (ed.) The Species Concept in Palaeontology. *Syst. Assoc. Publ.* **2**, 83–94.

Jung, W. W. (1968). *Hirmerella muensteri* (Schenk) Jung nov. comb., eine bedeutsame Konifere des Mesozoikums. *Palaeontographica*, *B***122**, 56–93.

Jung, W. W. (1974). Die Konifere *Brachyphyllum nepos* Saporta aus den Solnhofener Plattenkalk (unteres untertithon), ein Halophyt. *Mitt. Bayer. Staatssamml. Palaeont. hist. Geol.* **14**, 49–58, 2 pl.

Kedves, M. (1960). Etudes palynologiques dans le bassin de Dorog, 1. *Pollen et Spores*, **2**, 89–118.

Kelner-Pillault, S. (1969). Les abeilles fossiles. *Mem. Soc. Entomol. Ital.* **48**, 520–34.

Kemp, E. M. (1968). Probable Angiosperm pollen from the British Barremian to Albian Strata. *Palaeontology*, **11**, 421–34, 3 pl.

Kemp, E. M. (1970). Aptian and Albian miospores from Southern England. *Palaeontographica*, Stuttgart, **131** *B*, 73–143, 20 pl.

Kendall, M. W. (1948). On six species of *Pagiophyllum* from the Jurassic of Yorkshire and Southern England. *Ann. Mag. Nat. Hist.* (Ser 12), **1**, 73–108.

Kendall, M. W. (1949*a*). On a new Conifer from the Scottish Lias. *Ann. Mag. Nat. Hist.* (Ser. 12), **2**, 299–307.

Kendall, M. W. (1949*b*). On *Brachyphyllum expansum* (Sternberg) Seward, and its cone. *Ann. Mag. Nat. Hist.* (Ser. 12), **2**, 308–19.

Kendall, M. W. (1949*c*). A Jurassic Member of the Araucariaceae. *Ann. Botany* (London) N.S. **13**, 50, 151–61.

Kendall, M. W. (1952). Some conifers from the Jurassic of England. *Ann. Mag. Nat. Hist.* (Ser. 12), **5**, 583–94.

Knobloch, E. (1971). Neue Pflanzenfunde aus dem böhmischen und mährischen Cenoman. *N. Jb Geol. Paläont. Abh.* **139**, 43–56.

Koch, B. E. (1964). Review of Fossil Floras and Nonmarine Deposits of West Greenland. *Geol. Soc. America Bull.* **75**, 535–48, 1 pl.

Konijnenburg-van Cittert, J. H. A. van (1971). In situ gymnosperm pollen from the Middle Jurassic of Yorkshire. *Acta Bot. Neerl.* **20**, 1–96.

Krassilov, V. A. (1967). *The Early Cretaceous flora of the Southern Primorye and its significance for stratigraphy*. Moscow. (In Russian.)

Krassilov, V. A. (1968). On the classification of stomata. *Paleont. Zh.* **1**, 102–9, 1 pl. (In Russian.)

Krassilov, V. A. (1970). Approach to the classification of Mesozoic 'Ginkgoalean' plants from Siberia. *Palaeobotanist*, **18**, 12–19, 3 pl.

Krassilov, V. A. (1972*a*). Mesozoic flora of Burei (Ginkgoales and Czekanowskiales). *Far-East Geol. Inst. Acad. Sci. USSR, Moscow*, 150 pp., 34 pl. (In Russian.)

Krassilov, V. A. (1972*b*). Phytogeographical classification of Mesozoic floras and their bearing on continental drift. *Nature*, **237**, 49–50.

Krassilov, V. A. (1973*a*). The Jurassic disseminules with pappus and their bearing in the problem of angiosperm ancestry. *Geophytology*, **3**, 1–4.

Krassilov, V. A. (1973*b*). Upper Cretaceous staminate heads with pollen grains. *Palaeontology*, **16**, 41–4, 1 pl.

Krassilov, V. A. (1973*c*). Mesozoic plants and the problem of angiosperm ancestry. *Lethaia*, **6**, 163–78.

Krassilov, V. A. (1973*d*). Climatic changes in Eastern Asia as indicated by fossil floras, 1. Early Cretaceous. *Palaeogeogr. Palaeoclimatol. Palaeoecol.* **13**, 261–73.

Krassilov, V. A. (1973*e*). Cuticular structure of Cretaceous angiosperms from the Far East of the USSR. *Palaeontographica B***142**, 105–16.

Krassilov, V. A. (ed.) (1973*f*). Fossil floras and phytogeography of the Far East. *Far-East Geol. Inst. Acad. Sci. USSR, Vladivostok*. 115 pp., 26 pl. (In Russian.)

Krassilov, V. A. (1974). *Podocarpus* from the Upper Cretaceous of eastern Asia and its bearing on the theory of conifer evolution. *Palaeontology*, **17**, 365–70.

Kräusel, R. (1919). Die fossilen Koniferenhölzer. *Palaeontographica*, Stuttgart, **62**, 185–275.

Krutzsch, W. (1957). Sporenpaläontologische Untersuchungen in der sachsich-bohmischen Kreide und die Gliederung der Oberkreide auf mikrobotanischer Grundlage. *Ber. geol. Ges. DDR*, **2**, 123–9.

Kryshtofovich, A. (1918). On the Cretaceous flora of Russian Sakhalin. *J. Fac. Sci., Imperial Univ. Tokyo*, **11**, 1–73.

Kubitzki, K. (1969). Chemosystematischen Betrachtungen zur Grossgliederung der Dicotylen. *Taxon*, **18**, 360–68.

Laing, J. F. (1975*a*, in press). The stratigraphic setting of early angiosperm pollen. *Linn. Soc. Sympos.*, Academic Press.

Laing, J. F. (1975*b*, in press). Mid-Cretaceous angiosperm pollen from Southern England and Northern France. *Palaeontology*.

Lammons, J. M. (1970). *Pentaspis*, a new palynomorph genus from the Cretaceous (Aptian) of Peru. *Micropaleontology*, **16**, 175–8.

Langenheim, J. H. (1964). Present status of botanical studies of amber. *Harvard Univ. Bot. Mus. Leaflets*, **20**, 225–87, 3 pl.

Langenheim, J. H. (1969). Amber: a botanical enquiry. *Science*, **163**, 1157–69.

Langenheim, R. L., Jr., Smiley, C. J. and Gray, J. (1960). Cretaceous amber from the Arctic coastal plain of Alaska. *Geol. Soc. Amer. Bull.* **71**, 1345–56.

Lebedev, E. L. (1974). Albian flora and Lower Cretaceous stratigraphy of west Priokhotie. *Trans. Geol. Inst. Acad. Sci. USSR*, **254**, 1–147, 31 pl. (In Russian.)

Leppik, E. E. (1963). Reconstruction of a Cretaceous *Magnolia* flower: In Chandra, L. (ed.) *Advancing Frontiers of Plant Sciences*, vol. 4, Inst. Advancement Sci. Cult., New Delhi, pp. 79–94.

Leppik, E. E. (1971). Palaeontological evidence on the morphogenic development of flower types. *Phytomorphology*, **21**, 164–74.

References

Lesquereux, L. (1892). The Flora of the Dakota group. *US geol. Surv. Monogr.* 17.

Lobreau-Callen, D. (1972). Pollen des Icacinaceae, 1. Atlas (1). *Pollen et Spores*, **14**, 345–88, 16 pl.

Lowenstam, H. A. (1964). Palaeotemperatures of the Permian and Cretaceous periods. In Nairn, A. E. M. (ed.) *Problems in Palaeoclimatology*, Interscience, New York, pp. 227–52.

Lundblad, B. (1968). The present status of the genus *Pseudotorellia* Florin (Ginkgophyta). *J. Linn. Soc. Lond. Bot.* **61**, 189–95.

Maekawa, F. (1962). *Vojnovskaya* as a presumed ancestor of angiosperms. *J. Japan Bot.* **37**, 21–4.

Martin, A. R. H. (1968). *Aquilapollenites* in the British Isles. *Palaeontology*, **11**, 549–53, 7 pl.

Martin, A. R. H. (1973). Re-appraisal of some palynomorphs of supposed proteaceous affinity. *Spec. Publs geol. Soc. Aust.* **4**, 73–8, 1 pl.

May, F. E. and Traverse, A. (1973). Palynology of the Dakota Sandstone (Middle Cretaceous) near Bryce Canyon, National Park, Southern Utah. *Geoscience and Man*, **7**, 57–64, 1 pl.

McGinitie, H. D. (1941). A middle Eocene flora from the Sierra Nevada. *Publs Carnegie Instn*, **534**, 175 pp.

Médus, J. and Pons, A. (1967). Etude palynologique du Crétacé pyrénéo-provençal. *Rev. Palaeobotan. Palynol.* **2**, 111–17.

Meeuse, A. D. J. (1961). The Pentoxylales and the origin of monocotyledons. *Proc. Konink. Nederl. Akad. Wetensch. Amsterdam*, *C* **64**, 543–59.

Meeuse, A. D. J. (1970). The descent of the flowering plants in the light of new evidence from phytochemistry and from other sources. *Acta Bot. Neerl.* **19**, 61–72 and 133–40.

Mello, J. F. (1969). Paleontologic data storage and retrieval. In *Proc. N. Amer. Paleont. Convention*, *B*, pp. 57–71.

Melville, R. (1962). A new theory of the angiosperm flower. *Kew Bull.* **16** (1).

Metcalfe, C. R. and Chalk, L. (1950). *Anatomy of the Dicotyledons* (2 vols.), Clarendon, Oxford.

Millioud, M. E. (1967). Palynological study of the type localities at Valangin and Hauterive. *Rev. Palaeobotan. Palynol.* **5**, 155–67.

Muir, M. D. (1964). The palaeoecology of the small spores of the Middle Jurassic of Yorkshire. Ph.D. Thesis, University of London, 214 pp.

Muller, J. (1968). Palynology of the Pedawan and Plateau Sandstone Formations (Cretaceous – Eocene) in Sarawak, Malaysia. *Micropaleontology*, **14**, 1–37, 5 pl.

Muller, J. (1970). Palynological evidence on early differentiation of angiosperms. *Biol. Rev.* **45**, 417–50.

Muller, J. (1974). A comparison of Southeast Asian with European fossil angiosperm pollen floras. *Birbal Sahni Inst. Palaeobotan. Spec. Publ.* (Lucknow), **1**, 49–56.

Müller, H. (1966). Palynological investigations of Cretaceous sediments in northeastern Brazil. In Van Hinte, J. E. (ed.) *Proc. Z. West Afr. Micropalaeont. Colloq.* (Ibadan). Brill, Leiden, pp. 123–36.

Naumova, S. N. (1950). Pollen du type Angiosperme dans les depôts du Carbonifère inférieur. *Izv. Akad. Nauk. USSR*, Ser. géol. **3**, 103–13. (In Russian.)

Nemejc, F. (1956). On the problem of the origin and phylogenetic development of the Angiosperms. *Acta Mus. Nat. Prague*, **12***B*, 65–144.

Neves, R. (1958). Upper Carboniferous plant spore assemblages from the *Gastrioceras subcrenatum* horizon, north Staffordshire. *Geol. Mag.* **95**(1), 1–19.

Norris, G. (1967). Spores and pollen from the Lower Colorado Group (Albian–?Cenomanian) of central Alberta. *Palaeontographica*, Stuttgart, **120***B*, 72–115.

Oldham, T. C. B. (in press). The plant debris beds of the English Wealden. *Palaeontology*.

Orlov, Y. A. (1963). Gymnosperms and Angiosperms. *Osnovy Paleontologii*, 649 pp., 80 pls. (In Russian.)

Pacltová, B. (1961). Zur Frage der Gattung *Eucalyptus* in der böhmischen Kreideformation. *Preslia*, **33**, 113–29.

Pacltová, B. (1963). Derzeitiger Stand der paläobotanischen Erforschung der Kreidesedimente in Böhmen. *Ber. Geol. Ges. DDR*, **8**, 237–40.

Pacltová, B. (1966). Pollen grains of angiosperms in the Cenomanian Peruc Formation in Bohemia. *Palaeobotanist*, **15**, 52–4.

Paden Phillips, P. and Felix, C. J. (1971). A study of Lower and Middle Cretaceous spores and pollen from the southeastern United States; 1, Spores. *Pollen et Spores*, **13**, 447–73.

Pettitt, J. M. and Chaloner, W. G. (1964). The ultrastructure of the Mesozoic pollen *Classopollis*. *Pollen et Spores*, **6**, 611–20.

Pflug, H. D. (1953). Zur Entstehung und Entwicklung des Angiospermiden Pollens in der Erdgeschichte. *Palaontographica*, Stuttgart, *B* **95**, 60–171.

Playford, G. (1962–63). Lower Carboniferous microfloras of Spitsbergen. *Palaeontology*, **5**, 550–678, 18 pls.

Pocock, S. A. J. (1962). Microfloral analysis and age determination of strata at the Jurassic–Cretaceous boundaries in the western Canada plains. *Palaeontographica*, Stuttgart, *B* **111**, 1–95.

Potonié, R. (1960). Synopsis der Gattungen der Sporae Dispersae, 3. Teil. *Beih. Geol. Jb.* **39**, 1–189.

Potonié, R. (1973). 'Gattungen' der Sporae Dispersae ohne nomenklatorischen typus? *Grana*, **13**, 65–73.

Proctor, M. and Yeo, P. (1973). *The Pollination of Flowers*. Collins, London.

Reid, E. M. and Chandler, M. E. J. (1933). *The London Clay Flora*, Brit. Mus. Nat. Hist., London, 561 pp.

Reymanovna, M. (1968). On seeds containing *Eucommiidites troedssonii* pollen from the Jurassic of Grojec, Poland. *J. Linn. Soc. Lond. Bot.* **61**, 147–52, 1 pl.

Reyre, Y. (1970). Stereoscan observations on the pollen genus *Classopollis* Pflug 1953. *Palaeontology*, **13**, 303–22, 6 pls.

Rohdendorf, B. B. *et al.* (1968). *Jurassic insects of Karatau*. Akad. Nauk. SSSR (Division of General Biology), Moscow. 252 pp., 25 pl. (In Russian.)

Rouse, G. E. (1959). Plant microfossils from the Kootenay coal-measure strata of British Columbia. *Micropaleontology*, **5**, 303–24.

Sahni, B. (1932). A petrified *Williamsonia* (*W. Sewardiana* sp. nov.) from the Rajmahal Hills, India. *Mem. Geol. Surv. India, Pal. Indica*, N.S. **20**, Mem. 3.

Sahni, B. (1948). Palaeobotany in India, 6. *J. Indian Botan. Soc.* **26**, 241–73.

References

Samylina, V. A. (1959). New occurrences of angiosperms from the Lower Cretaceous of the Kolyma basin. *Bot. J.* **44**, 483–91. (Russian text with English summary.)

Samylina, V. A. (1960). Angiosperms from the Lower Cretaceous of the Kolyma basin. *Bot. J.* **45**, 335–52. (Russian text with English summary.)

Samylina, V. A. (1961). New data on the Lower Cretaceous flora of the southern part of the Maritime Territory of the RSFSR. *Bot. Zh. SSSR*, **46**, 634–45. (Russian with English summary.)

Samylina, V. A. (1968). Early Cretaceous angiosperms of the Soviet Union based on leaf and fruit remains. *J. Linn. Soc. Lond. Bot.* **61**, 207–18.

Samylina, V. A. (1974). Early Cretaceous floras of the Far East of the USSR. *Botan. Inst. Komarov Acad. Sci. USSR*, Leningrad (Komarov Chten. 27), 56 pp. (In Russian.)

Saporta, G. de (1894). *Flora fossile du Portugal.* Lisbon.

Sastri, R. N. L. (1969). Comparative morphology and phylogeny of the Ranales. *Biol. Rev.* **44**, 291–319.

Scheuring, B. W. (1970). Palynologische und palynostratigraphische Untersuchungen des Keupers im Bölchentunnel (Solothurner Jura). *Schweiz. Paläont. Abh.* **88**, 1–119, 43 pl.

Schlee, D. and Dietrich, H.-G. (1970). Insektführender Bernstein aus der Unterkreide des Libanon. *Neues Jb. Geol. Pälaont. Mh.* 1970, 40–50.

Schlinger, E. I. (1974). Continental Drift, *Nothofagus*, and some ecologically associated insects. *Ann. Rev. Entomol.* **19**, 323–43.

Schulz, E. (1967). Sporenpaläontologische Untersuchungen rätoliassischer Schichten im Zentrateil des Germanischen Beckens. *Paläontol. Abh.* Ser. 2, **3**, 427–633.

Scott, R. A. (1954). Fossil fruits and seeds from the Eocene Clarno formation of Oregon. *Palaeontographica*, **96***B*, 66–97.

Scott, R. A. (1960). Pollen of *Ephedra* from the Chinle formation (Upper Triassic) and the genus *Equisetosporites*. *Micropaleontology*, **6**, 271–6.

Scott, R. A. (1969). Silicified fruits and seeds from Western North America. *J. Paleont.* **43**, 898 (abstract).

Scott, R. A. and Barghoorn, E. S. (1958). *Phytocrene microcarpa* – a new species of Icacinaceae based on Cretaceous fruits from Kreischerville, New York. *Palaeobotanist*, **6**, 25–8.

Scott, R. A., Barghoorn, E. S. and Leopold, E. B. (1960). How old are the Angiosperms? *Amer. J. Sci.* **258***A*, 284–99.

Scott, R. A. *et al.* (1972). 'Pre-Cretaceous' angiosperms from Utah: Evidence for Tertiary ape of the palm woods and roots. *Amer. J. Bot.* **59**, 886–96.

Seward, A. C. (1895). *The Wealden Flora,* Vol. 2, Brit. Mus. Nat. Hist.

Seward, A. C. (1900). La flore Wealdienne de Bernissart. *Mém. Mus. Hist. Nat. Belg.* **1**, 37 pp., 4 pl.

Seward, A. C. (1913). A contribution to our knowledge of Wealden floras, with special references to a collection of plants from Sussex. *Q. Jl geol. Soc.* **69**, 85–116, 4 pl.

Seward, A. C. (1917). *Fossil Plants* Vol. 3, Cambridge University Press, London.

Seward, A. C. (1919). *Fossil Plants* Vol. 4, Cambridge University Press, London, 543 pp.

Seward, A. C. (1933). *Plant Life Through the Ages.* Cambridge University Press, London. 2nd edition.

Shaw, A. B. (1964). *Time in Stratigraphy*, McGraw-Hill, New York, 365 pp.
Simpson, G. G. (1960). Notes on the measurement of formal resemblance. *Amer. J. Sci.* **258***A*, 300–311.
Singh, C. (1964). Microflora of the Lower Cretaceous Mannville Group, east-central Alberta. *Bull. Res. Counc. Alberta*, **15**, 1–238, 29 pl.
Singh, C. (1971). Lower Cretaceous microfloras of the Peace River area, north-western Alberta. *Bull. Res. Counc. Alberta*, **28**, 542 pp., 8 pl.
Smart, J. and Hughes, N. F. (1972). The insect and the plant: progressive palaeoecological integration. In van Emden, H. F. Insect/plant relationships. *Sympos. roy. Entomol. Soc.* **6**, 143–55.
Smiley, C. J. (1969*a*). Cretaceous floras of Chandler–Colville region, Alaska: Stratigraphy and preliminary floristics. *Amer. Assoc. Petr. Geol. Bull.* **53**, 482–502.
Smiley, C. J. (1969*b*). Floral zones and correlations of Cretaceous Kukpowrak and Corwin Formations, North western Alaska. *Amer. Assoc. Petr. Geol. Bull.* **53**, 2079–93.
Smiley, C. J. (1970). Later Mesozoic flora from Maran, Pahang, West Malaysia. *Bull. Geol. Soc. Malaysia*, **3**, 77–113, 5 pl.
Smiley, C. J. (1972). Plant megafossil sequences, North Slope Cretaceous. *Geoscience and Man*, **4**, 91–9.
Smith, A. G. S., Briden, J. C. and Drewry, G. E. (1973). Phanerozoic world maps. In Hughes, N. F. (ed.) Organisms and continents through time. *Spec. Pap. Palaeont.* **12**, 1–42.
Sporne, K. R. (1965). *The Morphology of Gymnosperms*. Hutchinson, London.
Sporne, K. R. (1969). The ovule as an indicator of evolutionary status in angiosperms. *New Phytol.* **68**, 555–66.
Sporne, K. R. (1970). The advancement index and tropical rain-forest. *New Phytol.* **69**, 1171.
Sporne, K. R. (1972). Some observations on the evolution of pollen types in dicotyledons, *New Phytol.* **71**, 181–6.
Sporne, K. R. (1973). The survival of archaic dicotyledons in tropical rain-forests. *New Phytol.* **72**, 1175–84.
Sporne, K. R. (1974). Pollen evolution in dicotyledons. *Birbal Sahni Inst. Spec. Publ. Palaeobotan.*, Lucknow, **1**, 57–61.
Stebbins, G. L. (1972). Ecological distribution of centres of major adaptive radiation in the angiosperms. In Valentine, D. H. (ed.) *Taxonomy, Phytogeography and Evolution*. Academic Press, London.
Stopes, M. C. (1915). *Catalogue of Cretaceous plants in the British Museum (Nat. Hist.)*. Part 2.
Stover, L. E. (1963). Some Middle Cretaceous palynomorphs from West Africa. *Micropaleontology*, **9**, 85–94.
Takhtajan, A. L. (1969). *Flowering plants: Origin and Dispersal*. Oliver and Boyd, Edinburgh.
Teixeira, C. (1948). *Flora Mesozoica Portuguesa*. Vol. 1, Lisbon.
Teixeira, C. (1950). *Flora Mesozoica Portuguesa* Vol. 2, Lisbon.
Teixeira, C. (1954). La flore fossile des calcaires lithographiques de Santa Maria de Meya (Lérida, Espagne). *Bol. Soc. Geol. Portugal* **11**, 139–52.
Teslenko, Y. V. (1967). Some aspects of the evolution of terrestrial plants. *Geol. Geophys.*, Novosibirsk, **11**, 58–64. (In Russian.)

References

Teterjiuk, V. K. (1958). On finding open-pored pollen grains of Palaeozoic Angiosperms. *Dokl. Akad. Nauk. SSSR*, **118**, 1034–5.

Thomas, H. H. and Bancroft, N. (1913). On the cuticles of some Recent and Fossil Cycadean fronds. *Trans. Linn. Soc. Lond. Bot.* (Ser. 2), **8**, 155–205.

Thomas, H. H. (1925). The Caytoniales, a new group of Angiospermous plants from the Jurassic rocks of Yorkshire. *Philos. Trans.*, London, **213*B***, 299–363, 5 pl.

Tidwell, W. D. *et al.* (1970). *Palmoxylon simperi* and *Palmoxylon pristina*: two pre-Cretaceous angiosperms from Utah. *Science*, **168**, 835–40.

Tidwell, W. D., Rushforth, S. R. and Simper, A. D. (1970). Pre-Cretaceous flowering plants. Further evidence from Utah. *Science*, **170**, 547–8.

Tidwell, W. D., Rushforth, S. R. and Simper, A. D. (1971). *Rhizopalmoxylon* from the Arapien Shale, Jurassic of Utah. *Amer. J. Bot.* **58**, 473 (abstract).

Townrow, J. A. (1962). On *Pteruchus*, a microsporophyll of the Corystospermaceae. *Bull. Brit. Mus.* (*Nat. Hist.*) *Geol.* **6**, 287–320.

Townrow, J. A. (1967*a*). On *Rissikia* and *Mataia*, podocarpaceous conifers from the Lower Mesozoic of Southern Lands. *Pap. Proc. Roy. Soc. Tas.* **101**, 103–36.

Townrow, J. A. (1967*b*). On a conifer from the Jurassic of East Antarctica. *Pap. Proc. Roy. Soc. Tas.* **101**, 137–48.

Townrow, J. A. (1967*c*). The *Brachyphyllum crassum* complex of fossil conifers. *Pap. Proc. Roy. Soc. Tas.* **101**, 149–72.

Tralau, H. (1968). Botanical investigations into the fossil flora of Eriksdal in Fyledalen, Scania, 2. The Middle Jurassic microflora. *Sver. geol. Unders. C* **633**, 1–185, 26 pl.

Turutanova-Ketova, A. I. (1930). Jurassic flora of the Chain Kara Tau. *Trudy geol. Muz.* **6**, 131–72. (In Russian.)

Urey, H. C. (1973). Cometary collisions and geological periods. *Nature*, **242**, 32–3.

Vakhrameev, V. A. (1952). The stratigraphy and the fossil flora of the Cretaceous deposits of western Kazakhstan. In *Regional stratigraphy of the USSR*, vol. 1, Moscow. (In Russian.)

Vakhrameev, V. A. (1964). Jurassic and Early Cretaceous florae of Eurasia and the palaeofloristical provinces of that time. *Trudy geol. Inst. Akad. Nauk SSSR* (Geol. Ser.), **102**, 230–48. (In Russian.)

Vakhrameev, V. A. (ed.) (1968). Mesozoic plants. *Trudy geol. Inst. Akad. Nauk SSSR* (Geol. Ser.), Moscow, **191**, 88 pp., 19 pl. (In Russian.)

Vakhrameev, V. A. (1971). Development of the Early Cretaceous flora in Siberia. *Geophytology*, **1**, 75–83.

Vakhrameev, V. A., Dobrushkina, I. A., Zaklinskaya, E. D. and Meyen, S. V. (1970). Palaeozoic and Mesozoic floras of Eurasia and phytogeography of this time. *Trudy geol. Inst. Akad. Nauk SSSR* (Geol. Ser.), Moscow, **208**, 1–245. (In Russian.)

Vakhrameev, V. A. and Doludenko, M. P. (1961). Late Jurassic and early Cretaceous floras of the Bureja basin and their significance for stratigraphy. *Trudy geol. Inst. Akad. Nauk SSSR* (Geol. Ser.), Moscow, **59**, 135 pp., 60 pl. (In Russian.)

Velenovský, I. and Vinikłář, L. (1926–31). *Flora cretacea Bohemiae. Rozpr. Statn. Geol. Ustav. Cesk. Rep.* 1–345.

Vishnu-Mittre (1955). *Sporojuglandoidites jurassicus* gen. et sp.nov., a sporomorph from the Jurassic of the Rajmahal Hills, Bihar. *Palaeobotanist*, **4**, 151–2.

Ward, L. F., Fontaine, W. M., Bibbins, A., Wieland, G. R. (1905). Status of Mesozoic floras of the United States. *US geol. Surv. Monogr.* 48, 616 pp.

Watson, J. (1969). A revision of the English Wealden Flora, 1. Charales–Ginkgoales. *Bull. Brit. Mus. (Nat. Hist.) Geol.* **17**(5), 209–54, 6 pl.

Westwood, J. O. (1854). Contributions to fossil entomology. *Q. Jl geol. Soc. Lond.* **10**, 378–96, 5 pl.

Wieland, G. R. (1906 and 1916). *American Fossil Cycads*, 2 Vols, Washington.

Wieland, G. R. (1934). Fossil cycads, with special reference to *Raumeria reichenbachiana* Goeppert, of the Zwinger of Dresden. *Palaeontographica*, Stuttgart, **79**B, 85–130, 12 pl.

Yannin, B. T. (1958). Albian wood borers of the genus *Martesia* from the southern USSR. *Palaeont. J.* (AGI translation), **2**, 405–9.

Zaklinskaya, E. D. (1966). Pollen of angiosperms and its significance for the stratigraphy of the Upper Cretaceous and Palaeogene. *Trudy Inst. Geol. Nauk SSSR* (Geol. Ser.), Moscow, **74**, 258 pp. (In Russian.)

Zimina, V. G. (1967). First occurrence of *Vojnovskaya* in the Permian of South Maritime Territory. *Palaeont. J.* (AGI translation), **1**, 88–93.

Index

Index

Index

Index

Index

Index

Index

Index

Index